Plant Evolution and the Origin of Crop Species

Plant Evolution and the Origin of Crop Species

Editor

Ravikiran Vasant Mane

scitus academics

Plant Evolution and the Origin of Crop Species

Edited by **Ravikiran Vasant Mane**

Printed in 2017

ISBN: 978-1-68117-085-5

Library of Congress Control Number: 2015935484

SCITUS Academics LLC,
616, Corporate Way, Suite 2, 4766,
Valley Cottage, NY 10989

www.scitusacademics.com

Contents

Preface

Evolution is the force that shapes our living world. Countless different kinds of plants and animals pack the earth and each species is itself composed of a wide range of morphologies and adaptations. These species are continually being modified as they face the realities of their particular environments. In its simplest sense, evolution can be defined as a change in gene frequency over time. Genetic variability is produced by mutation and then that variability is shuffled and sorted by the various evolutionary forces. It does not matter whether the species are cultivated or wild, the basic evolutionary processes are the same. The way in which organisms evolve is dependent on their genetic characteristics and the type of environment they must face. A broad spectrum of evolutionary forces act to alter species including migration, selection and random chance. Over the last couple of decades, plant evolutionists have been using several classes of biochemical compounds to further assess levels of genetic variability and calculate evolutionary relationships. Alkaloids and flavonoids have achieved some popularity, but gel electrophoresis of enzymes has become the predominant mode of biochemical analysis. This technique takes advantage of the fact that proteins with different amino acid sequences often have different changes and physical conformations, and they migrate at different rates through a charged gel matrix. The major advantage of this type of analysis is that many loci and individuals can be measured simultaneously and most alleles are codominant so that heterozygotes can be identified.

Editor

Chapter 1

Evolution, Consequences and Future of Plant and Animal Domestication

Jared Diamond

Department of Physiology, University of California Medical School, Los Angeles, California 90095-1751, USA

ABSTRACT

Domestication interests us as the most momentous change in Holocene human history. Why did it operate on so few wild species, in so few geographic areas? Why did people adopt it at all, why did they adopt it when they did, and how did it spread? The answers to these questions determined the remaking of the modern world, as farmers spread at the expense of hunter–gatherers and of other farmers.

INTRODUCTION

Plant and animal domestication is the most important development in the past 13,000 years of human history. It interests all of us, scientists and non-scientists alike, because it provides most of our food today, it was prerequisite to the rise of civilization, and it transformed global demography. Because domestication ultimately yielded agents of conquest (for example, guns, germs and steel) but arose in only a few areas of the world, and in certain of those areas earlier than in others, the peoples who through biogeographic luck first acquired domesticates acquired enormous advantages over other peoples and expanded. As a result of those replacements, about 88% of all humans alive today speak some language belonging to one or another of a mere seven language families confined in the early Holocene to two small areas of Eurasia that happened to become the earliest centres of domestication — the Fertile Crescent and parts of China. Through that head start, the inhabitants of those two areas spread their languages and genes over much of the rest of the world. Those localized origins of domestication ultimately explain why this international journal of science is published in an Indo-European language rather than in Basque, Swahili, Quechua or Pitjantjatjara.

Much of this review is devoted to domestication itself: its origins, the biological changes involved, its surprising restriction to so few species, the restriction of its geographic origins to so few homelands, and its subsequent geographic expansion from those homelands. I then discuss the consequences of domestication for human societies, the origins of human infectious diseases, expansions of agricultural populations, and human evolution. After posing the unresolved questions that I would most like to see answered, I conclude by speculating about possible future domestications of plants and animals, and of ourselves. By a domesticate, I mean a species bred in captivity and thereby modified from its wild ancestors in ways making it more useful to humans who control its reproduction and (in the case of animals) its food supply. Domestication is thus distinct from mere taming of wild-born animals. Hannibal's African war elephants were, and modern Asian work elephants still are, just tamed wild individuals, not individuals of a genetically distinct population born and reared in captivity.

In 1997 I summarized available information about domestication and its consequences for human history in a book[1]. Since then, new details have continued to accumulate, and unanswered questions have come into sharper focus. Sources for statements not specifically referenced will generally be found in refs 1–9.

THE PAST OF DOMESTICATION

Our 'Decision' to Domesticate

The question "why farm?" strikes most of us modern humans as silly. Of course it is better to grow wheat and cows than to forage for roots and snails. But in reality, that perspective is flawed by hindsight. Food production could not possibly have arisen through a conscious decision, because the world's first farmers had around them no model of farming to observe, hence they could not have known that there was a goal of domestication to strive for, and could not have guessed the consequences that domestication would bring for them. If they had actually foreseen the consequences, they would surely have outlawed the first steps towards domestication, because the archaeological and ethnographic record throughout the world shows that the transition from hunting and gathering to farming eventually resulted in more work, lower adult stature, worse nutritional condition and heavier disease burdens[10, 11]. The only peoples who could make a conscious choice about becoming farmers were hunter–gatherers living adjacent to the first farming communities, and they generally disliked what they saw and rejected farming, for the good reasons just mentioned and others.

Instead, the origins of domestication involved unforeseen consequences of two sets of changes — changes in plants and animals, and changes in human behaviour. As initially recognized by Darwin[12], and elaborated by Rindos[13], many of the differences between domestic plants and their wild ancestors evolved as consequences of wild plants being selected, gathered and brought back to camp by hunter–gatherers, while the roots of animal domestication included the ubiquitous tendency of all peoples to try to tame or manage wild animals (including such unlikely candidates as ospreys, hyenas and grizzly bears). Although humans had been manipulating wild plants and animals for a long time, hunter–gatherer behaviour began to change at the end of the Pleistocene because of increasingly unpredictable climate, decreases in big-game species that were hunters' first-choice prey, and increasing human occupation of available habitats[14,15]. To decrease the risk of unpredictable variation in food supply, people broadened their diets (the so-called broad-spectrum revolution) to second- and third-choice foods, which included more small game, plus plant foods requiring much preparation, such as grinding, leaching and soaking[14, 16]. Eventually, people transported some wild plants (such as wild cereals) from their natural habitats to more productive habitats and began intentional cultivation[17].

The emerging agricultural lifestyle had to compete with the established hunter–gatherer lifestyle. Once domestication began to arise, the changes of plants and animals that followed automatically under domestication, and the competitive advantages that domestication conveyed upon the first farmers (despite their small stature and poor health), made the transition from the hunter–gatherer lifestyle to food production autocatalytic — but the speed of that transition varied considerably among regions[18, 19]. Thus, the real question about the origins of agriculture, which I consider below, is: why did food production eventually outcompete the hunter–gatherer lifestyle over almost the whole world, at the particular times and places that it did, but not at earlier times and other places?

Changes of Wild Species under Domestication

These changes are particularly well understood for southwest Asia's Fertile Crescent, the site of domestication that was earliest in the world and that yielded what are still the world's most valuable domestic plant and animal species. For most species domesticated there, the wild ancestor and its wild geographic range have been identified, its relation to the domesticate proven by genetic and chromosomal studies, its changes under domestication delineated (often at the gene level), those changes traced in successive layers of the archaeological record, and the approximate time and place of its domestication identified[9].

For example, wild wheats and barley bear their seeds on top of a stalk that spontaneously shatters, dropping the seeds to the ground where they can germinate (but where they also become difficult for humans to gather). An occasional single-gene mutation that prevents shattering is lethal in the wild (because the seeds fail to drop), but conveniently concentrates the seeds for human gatherers. Once people started harvesting those wild cereal seeds, bringing them back to camp, accidentally spilling some, and eventually planting others, seeds with a non-shattering mutation became unconsciously selected for rather than against[9, 17].

Individual wild animals also vary in traits affecting their desirability to humans. Chickens were selected to be larger, wild cattle (aurochs) to be smaller, and sheep to lose their bristly outer hairs (the kemp) and not to shed their soft inner hairs (the wool). Most domestic animals, including even recently domesticated trout[20], have smaller brains and less acute sense organs than do their wild ancestors. Good brains and keen eyes are essential to survival in the wild, but represent a quantitatively important waste of energy in the barnyard, as far as humans are concerned[3, 21].

Especially instructive are cases in which the same ancestral species became selected under domestication for alternative purposes, resulting in very different-appearing breeds or crops. For instance, dogs were variously selected to kill wolves, dig out rats, race, be eaten, or be cuddled in our laps. What naive zoologist glancing at wolfhounds, terriers, greyhounds, Mexican hairless dogs and chihuahuas would even guess them to belong to the same species? Similarly, cabbage (*Brassica oleracea*) was variously selected for its leaves (cabbage and kale), stems (kohlrabi), flower shoots (broccoli and cauliflower) and buds (brussels sprouts).

Why So Few Wild Species were domesticated

The wild animal species that most plausibly could have yielded valuable domesticates were large terrestrial mammalian herbivores and omnivores, of which the world holds 148 species weighing 45 kg or more (Table 9.2 of ref. 1). Yet only 14 of those 148 species were actually domesticated (Table 9.1 of ref. 1), prompting us to ask what prevented domestication of the other 134 species? Similarly, worldwide there are about 200,000 wild species of higher plants, of which only about 100 yielded valuable domesticates. Especially surprising are the many cases in which only one of a closely related group of species became domesticated. For example, horses and donkeys were domesticated, but none of the four zebra species congeneric and able to interbreed with them[3, 22].

The key question concerning this selectivity of domestication is as follows: in the cases of all those species never domesticated, did the difficulty lie with the species itself, or with the people indigenous to the area to which the species was native? For instance, is the abundance of large wild mammals the reason why no mammal species was ever domesticated in subequatorial Africa, making domestication superfluous for Africans? If that explanation were correct, then African people should also have ignored Eurasian domestic mammals when those were finally introduced to Africa, and European animal breeders on arriving in Africa should have succeeded in domesticating some African wild mammals, but both of those predictions are refuted by the actual course of history.

Six independent lines of evidence[1] converge to prove that, in most cases, the obstacle lay with the species itself, not with the local people: the rapid acceptance of introduced Eurasian domesticates by non-Eurasian peoples; the rapid ancient domestication of the most valuable wild species; the repeated independent domestications of many of them; the failure of even modern European plant and animal breeders to add significantly to our short list of valuable domesticates; ancient discoveries of the value of thousands of species that were regularly harvested in the wild but that never became domesticated; and the identification of the particular reasons preventing the domestication of many of those species.

Comparisons of domesticated wild species with never-domesticated close relatives illustrate the subtle factors that can derail domestication[1] (Fig. 1). For example, it is initially surprising that oak trees, the most important wild food plant in many parts of Eurasia and North America, were never domesticated. Like wild almonds, acorns of most individual wild oaks contain bitter poisons, with occasional non-poisonous mutant trees preferred by human foragers. However, the non-poisonous condition is controlled by a single dominant gene in almonds but polygenically in oaks, so that offspring of the occasional non-poisonous individuals are often non-poisonous in almonds but rarely so in oaks, preventing selection of edible oak varieties to this day. A second example is provided by the European horse breeders who settled in South Africa in the 1600s and — like African herders for previous millennia — tried to domesticate zebras. They gave up after several centuries for two reasons. First, zebras are incurably vicious, have the bad habit of biting a handler and not letting go until the handler is dead, and thereby injure more zoo-keepers each year than do tigers. Second, zebras have better peripheral vision than horses, making them impossible even for professional rodeo cowboys to lasso (they see the rope coming and flick away their head).

Figure 1: Comparisons of domesticated wild species (left of each pair) and their never-domesticated close relatives (right) reveal the subtle factors that can derail domestication.

Among wild mammal species that were never domesticated, the six main obstacles proved to be a diet not easily supplied by humans (hence no domestic anteaters), slow growth rate and long birth spacing (for example, elephants and gorillas), nasty disposition (grizzly bears and rhinoceroses), reluctance to breed in captivity (pandas and cheetahs), lack of follow-the-leader dominance hierarchies (bighorn sheep and antelope), and tendency to panic in enclosures or when faced with predators (gazelles and deer, except reindeer). Many species passed five of these six tests but were still not domesticated, because they failed a sixth test. Conclusions about non-domesticability from the fact of non-domestication are not circular, because these six obstacles can be assessed independently.

Why There Were So Few Homelands of Agriculture

Food production bestowed on farmers enormous demographic, technological, political and military advantages over neighbouring hunter–gatherers. The history of the past 13,000 years consists of tales of hunter–gatherer societies becoming driven out, infected, conquered or exterminated by farming societies in every area of the world suitable for farming. One might therefore have naively anticipated that, in any part of the world, one or more of the local hunter–gatherer societies would have stumbled upon domestication, become farmers, and thereby outcompeted the other local hunter–gatherer societies. In fact, food production arose independently in at most nine areas of the world (Fertile Crescent, China, Mesoamerica, Andes/Amazonia, eastern United States, Sahel, tropical West Africa, Ethiopia and New Guinea).

The puzzle increases when one scrutinizes that list of homelands. One might again naively have expected the areas most productive for farming today to correspond, at least roughly, to the areas most productive in the past. In reality, the list of homelands and the list of breadbaskets of the modern world are almost mutually exclusive (Fig. 2). The latter list includes California, North America's Great Plains, Europe, the pampas of Argentina, the cape of southern Africa, the Indian subcontinent, Java and Australia's wheat belt. Because these areas are evidently so well suited to farming or herding today, why were they not so in the past?

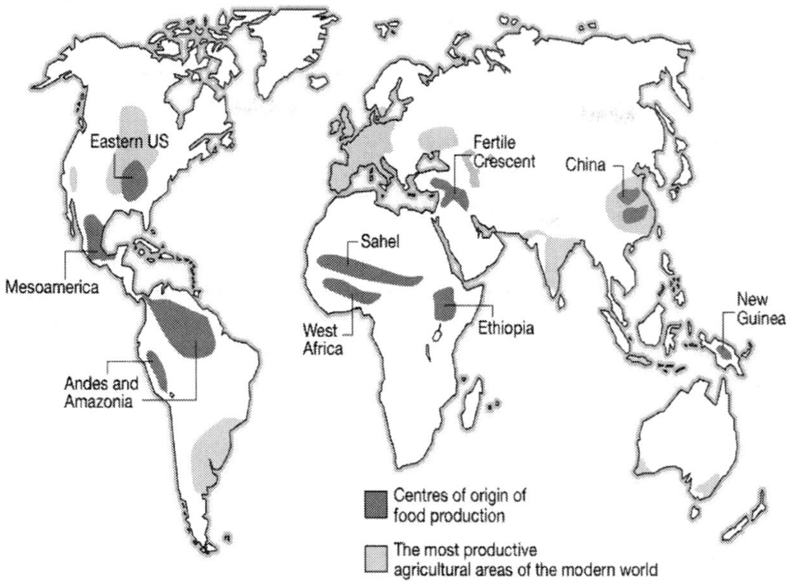

Figure 2: Ancient and modern centres of agriculture.

Ancient centres of origin of plant and animal domestication — the nine homelands of food production — are indicated by the orange-shaded areas on the map (based on Fig. 5.1 of ref.1). The most agriculturally productive areas of the modern world, as judged by cereals and major staples, are indicated by the yellow-shaded areas. Note that there is almost no overlap between the areas highlighted, except that China appears on both distributions, and that the most productive areas of the central United States today approach areas of the eastern United States where domestication originated. The reason why the two distributions are so different is that agriculture arose in areas to which the wild ancestors of the most valuable domesticable crops and animals were native, but other areas proved much more productive when those valuable domesticates reached them.

The explanation is that the homelands of agriculture were instead merely those regions to which the most numerous and most valuable domesticable wild plant and animal species were native. Only in those areas were incipient early farmers able to outcompete local hunter–gatherers. Once those locally available wild species had been domesticated and had spread outside the homelands, societies of homelands had no further advantage other than that of a head start, and they were eventually overtaken by societies of more fertile or climatically more favoured areas outside the homelands.

For instance, the Fertile Crescent of southwest Asia was home to wild wheats, barley, peas, sheep, goats, cows and pigs — a list that includes what are still the most valuable crops and livestock of the modern world. Hence hunter–gatherers of the Fertile Crescent domesticated those species and became the world's first farmers and herders, beginning around 8500 BC[1, 9, 23]. That head start in food production led to them and their close neighbours also developing the world's first metal tools, writing, empires and professional armies. Those tools of conquest, and Fertile Crescent human genes, gradually spread west into Europe and North Africa and east into the western Indian subcontinent and central Asia. However, once those crops, livestock and human inventions had spread, Fertile Crescent societies possessed no other advantages. As all of those elements slowly spread northwest across Europe, farming and power also shifted northwest from the Fertile Crescent to areas where farming had never arisen independently — first to Greece, then to Italy, and finally to northwest Europe. Human societies of the Fertile Crescent inadvertently committed slow ecological suicide in a zone of low rainfall prone to deforestation, soil erosion and salinization.

The Spread of Food Production

From the homelands of domestication, food production spread around the world in either of two ways. The much less common way was for hunter–gatherers outside the homelands to acquire crops or livestock from the homelands, enabling them to settle down as farmers or herders, as attested by archaeological evidence for substantial continuity of material culture, and by genetic, linguistic and skeletal evidence of continuity of human populations. The clearest such example of local adoption of food production is in southern Africa, where around 2,000 years ago some Khoisan hunter–gatherers acquired Eurasian livestock (cattle, sheep and goats) arriving from the north and became herders (so-called Hottentots). Much more often, however, local hunter–gatherers had no opportunity to acquire crops and livestock before they were overrun or replaced by farmers expanding out of the homelands, exploiting their demographic, technological, political and military advantages over the hunter–gatherers.

Expansions of crops, livestock, and even people and technologies tended to occur more rapidly along east–west axes than along north–south axes[1](Fig. 3). The reason is obvious: locations at the same latitude share identical day-lengths and seasonalities, often share similar climates, habitats and diseases, and hence require less evolutionary change or adaptation of domesticates, technologies and cultures than do locations at different latitudes. Examples include the rapid westwards and eastwards dispersal of wheat, horses, wheels and writing of western Asian origin, and the westwards dispersal of chickens, citrus and peaches of Chinese origin, along the east–west axis of Eurasia. This can be contrasted with the slow spread of Eurasian livestock and non-spread of Eurasian crops southwards along Africa's north–south axis[24], the slow spread of Mexican corn and the non-spread of Mexican writing and wheels and Andean llamas and potatoes along the Americas' north–south axis, and the slow spread of food production southwards along the north–south axis of the Indian subcontinent.

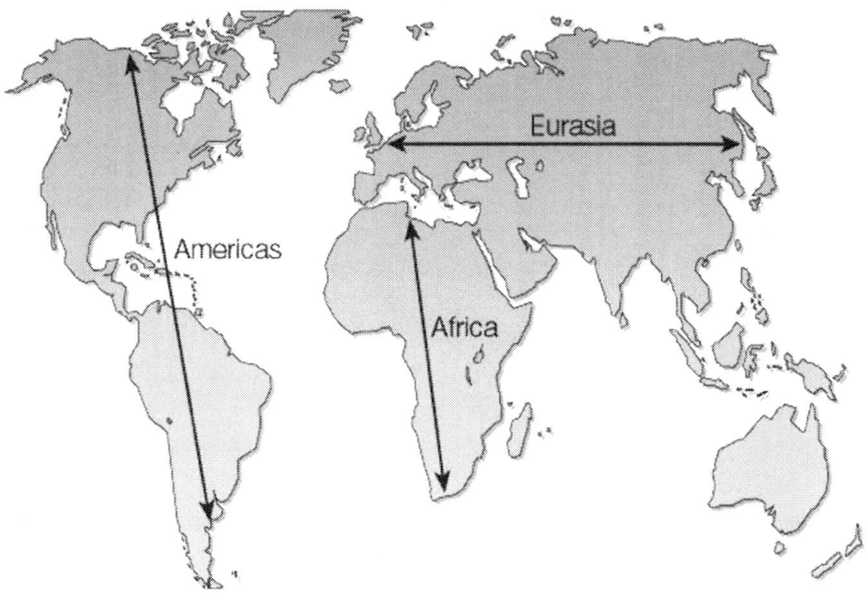

Figure 3: The continental major axis is oriented east–west for Eurasia but north–south for the Americas and Africa.

The spread of food production tended to occur more rapidly along east–west axes than along north–south axes, mainly because locations at the same latitudes required less evolutionary change or adaptation of domesticates than did locations at different latitudes. Modified from Fig. 10.1 of ref.1.

This is not to deny the existence of ecological barriers at the same latitude within Asia and North America, but the general pattern remains. Eurasia's east–west axis, and the resulting rapid enrichment of societies in each part of Eurasia by crops and technologies from other parts of Eurasia, became one of the main ultimate reasons why Eurasian peoples conquered Native American peoples, rather than visa versa. Eurasia's east–west axis also explains why there is much less evidence for multiple independent domestications of the same plant species (see below), and much more evidence for agriculturally driven language expansions, in Eurasia than in the Americas.

CONSEQUENCES OF DOMESTICATION

Consequences for Human Societies

Beginning around 8500 BC, the transition from the hunter–gatherer lifestyle to food production enabled people to settle down next to their permanent gardens, orchards and pastures, instead of migrating to follow seasonal shifts in wild food supplies. (Some hunter–gatherer societies in especially productive environments were also sedentary, but most were not). Food production was accompanied by a human population explosion that has continued unabated to this day, resulting from two separate factors. First, the sedentary lifestyle permitted shorter birth intervals. Nomadic hunter–gatherers had previously spaced out birth intervals at four years or more, because a mother shifting camp can carry only one infant or slow toddler. Second, plant and animal species that are edible to humans can be cultivated in much higher density in our gardens, orchards and pastures than in wild habitats.

Food production also led to an explosion of technology, because sedentary living permitted the accumulation of heavy technology (such as forges and printing presses) that nomadic hunter–gatherers could not carry, and because the storable food surpluses resulting from agriculture could be used to feed full-time craftspeople and inventors. By also feeding full-time kings, bureaucrats, nobles and soldiers, those food surpluses led to social stratification, political centralization and standing armies. All of these overwhelming advantages are what enabled farmers eventually to displace hunter–gatherers[1].

Evolution of Epidemic Infectious Diseases

The main killers of humans since the advent of agriculture have been acute, highly infectious, epidemic diseases that are confined to humans and that either kill the victim quickly or, if the victim recovers, immunize him/her for life[1, 25, 26, 27, 28]. Such diseases could not have existed before the origins of agriculture, because they can sustain themselves only in large dense populations that did not exist before agriculture, hence they are often termed 'crowd diseases'. The mystery of the origins of many of these diseases has been solved by molecular biological studies of recent decades, demonstrating that they evolved from similar epidemic diseases of our herd domestic animals with which we began to come into close contact 10,000 years ago. Thus, the evolution of these diseases depended on two separate roles of domestication: in creating much denser human populations, and in permitting much more frequent transmission of animal diseases from our domesticates than from hunted wild animals. For instance, measles and tuberculosis arose from diseases of cattle, influenza from a disease of pigs and ducks[1]. An outstanding mystery remains the origins of smallpox: did it reach us from camels or from cattle?

Crowd diseases paradoxically became agents of conquest, because exposed individuals acquired immune resistance from childhood exposure, and exposed populations gradually evolved genetic resistance, but unexposed populations had neither type of resistance. In practice, because 13 of our 14 large domestic mammals were Eurasian species, evolution of crowd diseases was concentrated in Eurasia, and the diseases became the most important agents by which Eurasian colonists expanding overseas killed indigenous peoples of the Americas, Australia, Pacific islands and southern Africa.

The Agricultural Expansions

Because some peoples acquired domesticates before other peoples could, and because domesticates conferred eventual advantages such as guns, germs and steel on the possessors, the history of the past 10,000 years has consisted of farmers replacing hunter–gatherers or less advanced farmers. These agricultural expansions, originating mainly from the nine homelands of agriculture, remade genetic and linguistic maps of the world (Table 18.2 of ref. 1). Among the most discussed (and often highly controversial) possible examples are the expansions of Bantu-speaking farmers out of tropical West Africa over subequatorial Africa[29], Austronesian-speaking farmers out of Taiwan over Island Southeast Asia[30], Fertile Crescent farmers over Europe[31, 32], and Korean farmers over Japan[33].

Human Genetic Evolution

Domestication has been by far the most important cause of changes in human gene frequencies in the past 10,000 years. Amosng the mechanisms responsible are: the spread of human genes from the agricultural homelands; the evolution of genetic resistance factors (including the ABO blood groups) to our new crowd infectious diseases[34, 35]; the evolution of adult-persistent lactase in milk-consuming populations of northern Europe and several parts of Africa; the evolution of allozymes of alcohol metabolism permitting consumption of large quantities of nutritionally important beer in western Eurasia; and the evolution of adaptations to a diet higher in simple carbohydrates, saturated fats and (in modern times) calories and salt, and lower in fibre, complex carbohydrates, calcium and unsaturated fats, than the hunter–gatherer diet[36].

UNSOLVED QUESTIONS

Among the host of unsolved questions, I focus here on six: what triggered the emergence of agriculture around 8500 BC and why did it not evolve earlier? Do crop and livestock species stem from a single domestication event or from multiple independent domestications? Can areas of food production be segregated into primary and secondary homelands, the latter describing areas where the arrival of primary homeland crops triggered local domestication? How did food production spread? Why were large domestic mammals predominantly Eurasian? And how can we gain a better understanding of the history of domestication of particular species?

Why Then But Not Earlier?

The human lineage diverged from that of chimpanzees around 6,000,000 years ago. For the next 99.8% of our separate history, there was no agriculture, until it emerged independently in up to nine areas on four continents in the short span of 6,000 years between 8500 and 2500 BC. All of those nearly-simultaneous independent origins seem to be too much of a coincidence. What triggered agriculture repeatedly then, and why had it never arisen during the previous 6,000,000 years?

Posing the question in this way both understates and overstates the puzzle. It understates the puzzle, because there are not only up to nine independent trajectories of intensification that did culminate in agriculture, but also many other ones that didn't quite (or that hadn't yet at the time that European conquest aborted them). Areas of the world where hunter–gatherers in the Holocene developed increased population densities, complex material culture, in some cases pottery, and (some anthropologists argue) sedentary living and ranked societies with chiefs included Mesolithic Europe, Japan and maritime Far East Asia, the North American high Arctic, the Pacific coast of northwest North America, interior California's oak woodlands, the California Channel Islands, the Calusa of Florida, the coast of Ecuador, and the Murray–Darling Basin of southeast Australia (for examples, see refs 37–39). But a similar intensification of hunter–gatherer societies also preceded the emergence of food production in its nine homelands; I suspect that the sole difference between the areas where people remained hunter–gatherers and the areas where food production evolved was that plant and animal species harvested in the latter but not the former areas included ones that automatically evolved domesticates, as already discussed. Thus, there were not just 5–9, but several dozen, independent trajectories of intensification in the Holocene.

On the other hand, my formulation of the question also overstates the puzzle. Only behaviourally modern *Homo sapiens* was biologically and mentally capable of the technological advances and foraging efficiency that resulted in intensified hunting and gathering, and (sometimes) in food production[40]. But behaviourally modern *Homo sapiens* did not emerge until around 55,000–80,000 years ago (the exact date is debated), so we should say that the independent simultaneous emergences were not concentrated in the last 0.2% of hominid history, but 'only' in the last 15% of modern human history. Still, even that seems too concentrated a bout of simultaneous emergences to be coincidental. Was it just that the origins of behaviourally modern *Homo sapiens* set clocks ticking by chance at the same rate all over the globe? That strains credulity, especially as intensified hunter–gatherer economies failed to arise in more areas than the areas in which they did arise.

A possible explanation seems to me to derive from four developments in the Late Pleistocene that may indeed have driven the clock's ticking. First, improvements in human hunting skills and consequent depletion or extermination of large mammalian prey would have made the hunter–gatherer lifestyle less rewarding and less able to compete with food production. Second was the development of human technology to collect, process and store wild foods (such as wild cereals), without which subsequently exploiting the same food species as domesticates would have been impossible (that is, what is the point of sowing wheat if you have not yet determined how to reap, roast and store it?). The third development was the on-going competition between human societies, such that those societies with more effective technology at any moment prevailed over other societies. Fourth, the gradual rise in human population numbers through the Pleistocene required intensified food procurement to feed those larger populations.

Against that background of gradual change, a trigger that may have caused intensification and food production to emerge only after the end of the Pleistocene would have been the end-of-Pleistocene climate changes in temperature, rainfall and unpredictability. These changes could have triggered the broad-spectrum reduction in diet[14, 15, 16, 17], and made agriculture possible in areas where it would have been impossible during the Ice Ages (for example, expanding Fertile Crescent woodland habitats with understories of wild cereals[41]). Once food production had thus begun, the autocatalytic nature of the many changes accompanying domestication (for example, more food stimulating population growth that required still more food) made the transition rapid. By this interpretation, the independent emergences of food production are no longer remarkably simultaneous — they could not have happened before the end of the Pleistocene (11000 BC), and after the end of the Pleistocene they occurred at very different times, ranging from about 8500 BC (in the Fertile Crescent) to about 2500 BC (eastern North America). Most of the links in this speculative hypothesis are in obvious need of testing.

Multiple versus Single Domestications

A long-standing question concerns whether each crop and livestock species stems from a single domestication event within a restricted geographic area, or from multiple independent domestications at different sites. An accumulation of recent evidence suggests to me the following generalization: that the former interpretation applies to most major Eurasian crops, the latter interpretation to many New World crops and the major Eurasian livestock species.

Among New World crops, many are represented by distinct but related species in South America, Mesoamerica and the eastern United States, leaving no doubt that related species were domesticated independently in these areas (for example, beans, chenopods, chilli peppers, cotton, squashes, tobaccos and possibly amaranths). Multiple independent domestications are attested within the same species for the chilli pepper species *Capsicum annuum*, common bean *Phaseolus vulgaris*, lima bean*Phaseolus lunatus* and squash species *Cucurbita pepo*[4, 7, 8, 42]. Conversely, the eight crops that founded Fertile Crescent agriculture, with the possible exception of barley, each seem to derive from only a single domestication event[5, 9, 43, 44, 45].

Evidence for separate independent domestications in western and more eastern parts of Eurasia are now available for all of the 'big five' domesticated mammals (cow, sheep, goat, pig and horse), plus one of the 'minor nine' (water buffalo)[46, 47, 48, 49, 50, 51, 52, 53, 54]. For example, cows were domesticated independently in the Fertile Crescent (yielding modern humpless cows), in the Indian subcontinent (yielding modern humped Zebu cows) and in North Africa[46, 50, 53].

I suggest the following hypothesis to explain predominantly single domestications of Fertile Crescent founder crops, but multiple domestications of Eurasian livestock and many New World crops. Except for barley and flax, the wild ancestors of the Fertile Crescent founder crops had restricted geographic ranges confined to the area between modern Turkey and western Iran, while chickpea was even more narrowly restricted, to southeastern Turkey. Those small geographic ranges, plus the rapid spread of domesticates along Eurasia's east–west axis, meant that, once a wild plant had been domesticated, it spread so rapidly that further independent domestications of the same or related species were pre-empted. The large Eurasian mammals, however, had such wide geographic ranges (in the case of pigs extending for 13,000 km from Spain to China) that there was ample time for independent domestications at locations west and east of each other. In the New World, even though all the homelands of agriculture lay within only 4000 km of each other, the slowness of crop diffusion along the New World's north–south axis meant that repeated independent domestications were frequent. So slow was that diffusion that the New World's main animal domesticates — the llama and guinea pig of the Andes, and the turkey of Mexico — had not even spread the mere 2,000 km north to Mexico and south to the Andes, respectively, by the time that Europeans arrived in AD 1492.

Primary versus Secondary Homelands

In several parts of the world, food production arose only upon the arrival of domesticates from the primary homelands, whereupon people proceeded to domesticate some local wild plants or animals that had not been domesticated previously[9]. Clear examples of such 'secondary' homelands, in which local domestication was triggered by the arrival of Fertile Crescent crops, were Europe (local domestication of poppies and possibly oats) and Egypt (chufa and sycamore fig).

The recognition of those secondary homelands requires us to reconsider the supposed primary homelands. On the one hand, some of the primary homelands may better be viewed as consisting of multiple homelands in which distinct systems of food production arose nearby but independently of each other. This is especially true for the homeland of Andes/Amazonia, which actually comprised primary highland sites in the Andes as well as primary lowland sites scattered from Panama through the Amazon Basin to the Pacific coast of Ecuador and Peru[55, 56]. Similarly, the Mesoamerican and Fertile Crescent homelands may have consisted of a mixture of highland and lowland sites, while China probably included northern and southern sites in the Yellow River and Yangtze River basins, respectively, as well as coastal lowland and interior upland sites.

On the other hand, some of the nine candidates for primary homelands may actually be secondary homelands in which domestication was triggered by the arrival of domesticates or of farmers from elsewhere. Independent origins of food production seem indisputable for five of the candidates (the Fertile Crescent, China, Mesoamerica, South America and eastern United States), because they were the earliest sites of domestication in their respective parts of the world. But questions have been raised, at least in conversation, regarding the independence of the other four candidates. Especially uncertain is the status of Ethiopia, where it is unknown whether several undoubted local domesticates (teff, coffee, finger millet, chat, noog and ensete) were cultivated before or only after the arrival of Fertile Crescent domesticates, and the New Guinea highlands, where remains of irrigation and drainage systems attest to early agriculture but where the first crops grown remain unidentified and the earliest dates of food production remain disputed. The independence of even the eastern United States has been challenged recently[42, 57], but the evidence seems compelling that Mexican crops arrived there only by way of southwestern United States and only long after local eastern origins of domestication[8,58]. Conversely, in southern India the exact dates of arrival of Fertile Crescent domesticates and of earliest cultivation of local domesticates remain uncertain.

Mechanism of the Spread of Food Production

As already noted, the spread of agriculture from its homelands involved in a few cases the acquisition of domesticates by hunter–gatherers outside the homelands, and in more cases the spread of farmers themselves from the homelands. The contributions of these two processes await resolution in many other cases. For example, contrary to what I wrote five years ago[1], the spread of farming in coastal west Mediterranean Europe (in the form of the Cardial and impressed ware cultures) now seems to have involved the rapid transport by sea of a complete package of Neolithic domesticates around 5400 BC by colonizing pioneer farmers[59]. The Yayoi horizon, which marks the arrival of intensive rice agriculture in Japan, and which Japanese scholars until recently preferred to view as an adoption of mainland practices by the indigenous pre-existing Japanese population, now seems increasingly likely on genetic evidence to represent the arrival, population increase and spread of Korean farmers[33].

Why Large Domestic Mammals Were Mainly Eurasian

Part of the reason why large domestic mammals were mainly Eurasian is simply that Eurasia, being the largest continent and having escaped the Late-Pleistocene extinctions that eliminated most large mammal species of the Americas and Australia[60], has the largest number of large wild mammal species. But there is a second part to the answer — a much higher percentage of large mammal species proved domesticable in Eurasia (18%) than in any other continent (Table 9.2 of ref. 1). Especially striking is the contrast of Eurasia with sub-Saharan Africa, where none of the 51 large mammal species was domesticable.

This difference constitutes a problem not in human behaviour, but in animal behaviour and sociobiology — something about African environments selected for one or more of the six mammalian traits that made domestication difficult. We already have some clues, as many of Africa's large mammals are species of antelopes and other open-country mammals whose herds lack the follow-the-leader dominance hierarchies characterizing Eurasian cattle, sheep, goats and horses[3, 61]. To resolve this problem, I suggest attempting to assign one or more of the six traits derailing domestication to each of the non-domesticated large mammal species of Eurasia and Africa, then evaluating the environmental factors behind the evolution of that trait.

History of Domestication of Particular Species

The history of domestication is much better understood for domesticates of western Eurasia than of other parts of world. Taking Zohary & Hopf's[9]account of western Eurasian plant domestication as a gold standard, it will be a challenge to workers on other biotas to match that standard. Even for western Eurasia, important unanswered questions abound. To mention only one out of dozens, calculation of molecular divergence times between dogs and wolves suggests that domestication of wolves began around 100,000 years ago[62, 63], yet the marked morphological differences between wolves and dogs (which should be easily detectable in fossilized skeletons) do not appear until about 11,000 years ago. How can the molecular data and the morphological data be reconciled?

THE FUTURE OF DOMESTICATION

Further Domestications of Plants and Animals

We humans today depend for our survival on that tiny fraction of wild species that has been domesticated. Might the rise of molecular biology, genetics and understanding of animal behaviour help feed our growing numbers by increasing that tiny fraction? Modern science has indeed made it technically possible to 'domesticate' species undomesticable in the past, in the sense that we achieve far more draconian control over the captive breeding of endangered California condors (computer-matched for mating to maximize genetic diversity) than the low-tech control that ancient animal breeders exerted over their livestock. But although this 'domestication' is of great interest to conservation biologists, it holds no promise of a condor industry to displace chicken from the supermarkets. What wild species might now be domesticated with profit?

It is instructive to reflect on the meagre additions to our repertoire of domestic species in recent millennia, despite monumental efforts. Of the world's 14 valuable big domestic mammals, the sole addition within the last millennium has been the reindeer, one of the least valuable of the 14. (In contrast, the five most valuable — the sheep, goat, cow, pig and horse — had all been domesticated repeatedly by 4000 BC.) Long-ongoing efforts by modern livestock breeders to domesticate other large wild mammals have resulted either in virtual failure (for example, eland, elk, moose, musk ox and zebra), or else in ranched animals (deer and American bison) that still cannot be herded and that remain of trivial economic value compared to the five most valuable mammals. Instead, all of the mammalian species that have recently become well established as domesticates (for example, arctic fox, chinchilla, hamster, laboratory rat and rabbit) are small mammals dwarfed in usefulness as well as in size by cows and sheep. Similarly, whereas several wild plants were first domesticated only in modern times (for example, blueberries, macadamia nuts, pecans and strawberries), their value is insignificant compared to that of ancient domesticates such as wheat and rice.

Our best hopes for valuable new domesticates lie in recognizing the specific difficulties that previously derailed domestication of particular valuable wild species, and using modern science to overcome those difficulties. For instance, now that we understand the polygenic control of non-bitterness in acorns, perhaps we could use that knowledge to select for oaks with non-bitter acorns, just as ancient farmers selected for non-bitterness controlled by a single gene in almonds. I am concerned, however, that such attempts may in the long run do us more harm than good. Humanity's greatest risk today is of our growing numbers and aspirations ultimately destroying our society by destroying our environment. Providing undernourished people with more food would be a laudable goal if it were inexorably linked to reducing our numbers, but in the past more food has always resulted in more people. Only when crop and animal breeders take the lead in reducing our numbers and our impacts will they end up by doing us net good.

Further Domestication of Humans

Some genotypes that used to serve us well as hunter–gatherers now serve us poorly as first-world citizens who forage only in supermarkets — especially metabolically thrifty genotypes that now predispose to type II diabetes, salt-conserving genotypes that predispose to hypertension, and other genotypes predisposing to other cardiovascular diseases and lipid disorders. As formerly spartan populations become westernized ('coca-colonized')[64], they fall victim to these diseases of the western lifestyle, extreme examples being the 70% incidence of type II diabetes in those Nauru Islanders and Pima Indians lucky enough to survive to the age of 60 (ref. 65). Because diabetes now afflicts south Asians and Pacific Islanders already in their twenties with high morbidity and mortality, there has been detectable natural selection against the predisposing genotypes even within just recent decades. The lower frequency of type II diabetes in Europeans than in non-Europeans matched for diet and lifestyle suggests that natural selection had already reduced European frequencies of those genotypes in previous centuries, as the western lifestyle was developing in Europe. In effect, the unconscious domestication of humans by agriculture that began over 10,000 years ago is still underway.

Even more such gene-frequency changes, also known as illness and deaths, are expected in the near future, as westernization accelerates in the world's two most populous countries, China and India[66, 67]. For example, the incidence of type II diabetes in mainland China, until recently less than 1%, has already tripled in some areas. What lies ahead for China can be projected by considering overseas Chinese populations in Hong Kong, Taiwan, Singapore and Mauritius, where westernization is further advanced and the incidence of type II diabetes is up to 17%. Similarly, the incidence in overseas Indian populations such as that of Fiji gives a foretaste of diabetes' future in India itself.

The resulting projections are that the number of cases of diabetes is expected to increase worldwide by 46% from the year 2000 to 2010, to reach around 220 million in 2010 and around 300 million in 2025. The steepest increase will be in east Asia (including China and India), the projected home of 60% of the world's diabetics in 2010. Similar diet-related disease epidemics are underway in less numerous peoples (from Africans to Aboriginal Australians), involving not just diabetes but also hypertension and other conditions. Thus, these epidemics pose the same dilemma as do efforts to domesticate more wild plant and animal species: how can we ensure that agriculture spreads only happiness, and not suffering as well?

ACKNOWLEDGEMENTS

It is a pleasure to acknowledge my debt to P. Bellwood, A. Ehrlich, K. Flannery, I. Hodder and B. Smith for valuable suggestions.

REFERENCES

1. Diamond, J. *Guns, Germs, and Steel: the Fates of Human Societies*(Norton, New York, 1997).

2. Cavalli-Sforza, L. L., Menozzi, P. & Piazza, A. *The History and Geography of Human Genes* (Princeton Univ. Press, Princeton, 1994).
3. Clutton-Brock, J. *Domesticated Animals From Early Times* (Univ. Texas Press, Austin, 1981).
4. Kiple, K. F. & Ornelas, K. C. *The Cambridge World History of Food*(Cambridge Univ. Press, Cambridge, 2000).
5. Ladizinsky, G. *Plant Evolution under Domestication* (Kluwer, Dordrecht, 1998).
6. Ruhlen, M. *A Guide to the World's Languages* (Stanford Univ. Press, Stanford, 1997).
7. Smartt, J. & Simmonds, N. W. *Evolution of Crop Plants* 2nd edn (Longman, Harlow, 1995).
8. Smith, B. D. *The Emergence of Agriculture* (Scientific American Library, New York, 1998).
9. Zohary, D. & Hopf, M. *Domestication of Plants in the Old World* 3rd edn (Oxford Univ. Press, Oxford, 2000).
10. Cohen, M. N. & Armelagos, G. J. *Paleopathology at the Origins of Agriculture* (Academic, Orlando, 1984).
11. Diamond, J. *The Third Chimpanzee: the Evolution and Future of the Human Animal* (HarperCollins, New York, 1992).
12. Darwin, C. *The Origin of Species by Means of Natural Selection*(Murray, London, 1859).
13. Rindos, D. *The Origins of Agriculture: An Evolutionary Perspective*(Academic, San Diego, 1984).
14. Flannery, K. V. in *The Domestication of Plants and Animals* (eds Ucko, P. J. & Dimbleby, G. W.) 73–100 (Ducksworth, London, 1969).
15. Binford, L. F. in *New Perspectives in Archaeology* (eds Binford, S. R. & Binford, L. R.) 313–341 (Aldine, Chicago, 1968).
16. Stiner, M. C., Munro, N. D., & Surovell, T. A. The tortoise and the hare: small-game use, the broad-spectrum revolution, and paleolithic demography. *Curr. Anthropol.* 41, 39–73 (2000).
17. Hillman, G. C., & Davies, M. S. Measured domestication rates in wild wheats and barley under primitive cultivation, and their archaeological implications. *J. World Prehist.* 4, 157–222 (1990).

18. Smith, B. D. Low-level food production. *J. Archaeol. Res.* 9, 1–43 (2001).
19. Smith, B. D. Documenting plant domestication: the consilience of biological and archaeological approaches. *Proc. Natl Acad. Sci. USA*98, 1324–1326 (2001).
20. Marchetti, M. & Nevitt, G. Effect of hatchery rearing on brain structures of rainbow trout, *Oncorhynchus mykiss. Env. Biol. Fishes*(in the press).
21. Diamond, J. Competition for brain space. *Nature* 382, 756–757 (1996).
22. Clutton-Brock, J. *Horse Power* (Harvard Univ. Press, Cambridge,1992).
23. Lev-Yadun, S., Gopher, A., & Abbo, S. The cradle of agriculture. *Science* 288, 1602–1603 (2000).
24. Marshall, F. in *The Origins and Development of African Livestock* (eds Blench, B. M. & MacDonald, K. C.) 191–221 (UCL Press, London,2000).
25. McNeill, W. *Plagues and Peoples* (Doubleday, Garden City, 1976).
26. Crosby, A. *The Columbian Exchange: Biological Consequences of 1492* (Greenwood, Westport, 1972).
27. Crosby, A. *Ecological Imperialism: The Biological Expansion of Europe, 900–1900* (Cambridge Univ. Press, Cambridge, 1986).
28. Ramenofsky, A. *Vectors of Death* (Univ. New Mexico Press, Albuquerque, 1987).
29. Ehret, C. *An African Classical Age* (Univ. Press of Virginia, Charlottesville, 1998).
30. Bellwood, P. *Prehistory of the Indo-Malaysian Archipelago* revised edn (Univ. Hawaii Press, Honolulu, 1997).
31. Ammerman, A. J. & Cavalli-Sforza, L. L. *Neolithic Transition and the Genetics of Populations in Europe* (Princeton Univ. Press, Princeton,1984).
32. Renfrew, C. *Archaeology and Language: The Puzzle of Indo-European Origins* (Cambridge Univ. Press, New York, 1987).
33. Hudson, M. *Ruins of Identity* (Univ. Hawaii Press, Honolulu, 1999).

34. Vogel, F. & Chakravartti, M. R. ABO blood groups and smallpox in a rural population of West Bengal and Bihar (India). *Hum. Genet.* 3, 166–180 (1966).

35. Diamond, J. & Rotter, J. I. in *The Genetic Basis of Common Diseases*(eds King, R. A., Rotter, J. I. & Motulsky, A. G.) 50–64 (Oxford Univ. Press, New York, 2002).

36. Eaton, S. B., Shostak, M. & Konner, M. *The Paleolithic Prescription*(Harper and Row, New York, 1988).

37. Bonsall, C. *The Mesolithic in Europe* (John Donald, Edinburgh, 1990).

38. Lourandos, *Continent of Hunter-Gathers: New Perspectives in Australian Prehistory* (Cambridge Univ. Press, Cambridge, 1997).

39. Arnold, J. E. *The Origins of a Pacific Coast Chiefdom: The Chumash of the Channel Islands* (Univ. Utah Press, Salt Lake City, 2001).

40. Klein, R. G. *The Human Career: Human Biological and Cultural Origins*2nd edn (Univ. Chicago Press, Chicago, 1999).

41. Richerson, P. F., Boyd, R. & Bettinger, R. L. Was agriculture impossible during the Pleistocene but mandatory during the Holocene? A climate change hypothesis. *Am. Antiquity* 66, 387–412 (2001).

42. Sanjur, O. I., Piperno, D. R., Andres, T. C. & Wessel-Beaver, L.Phylogenetic relationships among domesticated and wild species of*Cucurbita* (Cucurbitaceae) inferred from a mitochondrial gene: implications for crop plant evolution and areas of origin. *Proc. Natl Acad. Sci. USA* 99, 535–540 (2002).

43. Zohary, D. in *The Origins and Spread of Agriculture and Pastoralism in Eurasia* (ed. Harris, D. R.) 142–158 (UCL Press, London, 1996).

44. Zohary, D. Monophyletic vs. polyphyletic origin of the crops on which agriculture was founded in the Near East. *Genet. Resources Crop Evol.* 46, 133–142 (1999).

45. Heun, M., Schäfer-Pregl, R., Klawan, D. *et al.* Site of Einkorn wheat domestication identified by DNA fingerprinting. *Science* 278, 1312–1314 (1997).

46. Loftus, R. T. *et al.* Evidence for two independent domestications of cattle. *Proc. Natl Acad. Sci. USA* 91, 2757–2761 (1994).

47. Lau, C. H. *et al*. Genetic diversity of Asian water buffalo (*Bubalus bubalis*): mitochondrial DNA D-loop and cytochrome b sequence variation. *Anim. Genet.* 29, 253–264 (1998).
48. Vilà, C. *et al.* Widespread origins of domestic horse lineages. *Science*291, 474–477 (2001).
49. MacHugh, D. E. & Bradley, D. G. Livestock genetic origins: goats buck the trend. *Proc. Natl Acad. Sci. USA* 98, 5382–5384 (2001).
50. Troy, C. S. *et al.* Genetic evidence for Near Eastern origins of European cattle. *Nature* 410, 1088–1091 (2001).
51. Luikart, G. *et al.* Multiple maternal origins and weak phylogeographic structure in domestic goats. *Proc. Natl Acad. Sci. USA* 98, 5927–5932 (2001).
52. Giuffra, E. *et al.* The origin of the domestic pig: independent domestication and subsequent introgression. *Genetics* 154, 1785–1791 (2000).
53. Hanotte, O. *et al*. African pastoralism: genetic imprints of origins and migrations. *Science* 296, 336–339 (2002).
54. Hiendleder, S. *et al.* Molecular analysis of wild and domestic sheep questions current nomenclature and provides evidence for domestication from two different subspecies. *Proc. R.. Soc. Lond. B*269, 893–904 (2002).
55. Piperno, D. R. & Pearsall, D. M. *The Origins of Agriculture in the Lowland Neotropics* (Academic, San Diego, 1998).
56. Piperno, D. R., Rinere, A. J. Holst, I. & Hansell, P. Starch grains reveal early root crop horticulture in the Panamanian tropical forest.*Nature* 407, 894–897 (2000).
57. Lentz, D. *et al.* Prehistoric sunflower (*Helianthus annuus* L.) domestication in Mexico. *Econ. Bot.* 55, 370–376 (2001).
58. Smith, B. D. *Rivers of Change: Essays on Early Agriculture in Eastern North America* (Smithsonian Institution Press, Washington DC,1992).
59. Zilhão, J. Radiocarbon evidence for maritime pioneer colonization at the origins of farming in West Mediterranean Europe. *Proc. Natl Acad. Sci. USA* 98, 14180–14185 (2001).
60. Martin, P. S., & Klein, R. G. *Quaternary Extinctions* (Univ. Arizona Press, Tucson, 1984).

61. Spinnage, C. *The Natural History of Antelopes* (Fax on File, New York,1986).
62. Vilà, C. *et al.* Multiple and ancient origins of the domestic dog. *Science* 276, 1687–1689 (1997).
63. Vilà, C., Maldonado, J. E. & Wayne, R. K. Phylogenetic relationships, evolution, and genetic diversity of the domestic dog. *J. Hered.* 90, 71–77 (1999).
64. Zimmet, P. Globalization, coca-colonization and the chronic disease epidemic: can the Doomsday scenario be averted? *J. Intern. Med.*247, 301–310 (2001).
65. Dowse, G. K., Zimmet, P. Z., Finch, C. F. & Collins, V. R. Decline in incidence of epidemic glucose intolerance in Nauruans: implications for the "thrifty genotype". *Am. J. Epidemiol.* 133, 1093–1104 (1991).
66. Amos, A., McCarthy, D. & Zimmet, P. The rising global burden of diabetes and its complications: estimates and projections to the year 2010. *Diabetic Med.* 14, S1–85 (1997).
67. Zimmet, P., Alberti, K. G. M. M. & Shaw, J. Global and societal implications of the diabetes epidemic. *Nature* 414, 782–787 (2001).

Chapter 2

Evolutionary Origins and Dynamics of Octoploid Strawberry Subgenomes Revealed by Dense Targeted Capture Linkage Maps

Jacob A. Tennessen[1], Rajanikanth Govindarajulu[2], Tia-Lynn Ashman[2], and Aaron Liston[3]

[1]Department of Integrative Biology, Oregon State University
[2]Department of Biological Sciences, University of Pittsburgh
[3]Department of Botany and Plant Pathology, Oregon State University

ABSTRACT

Whole-genome duplications are radical evolutionary events that have driven speciation and adaptation in many taxa. Higher-order polyploids have complex histories often including interspecific hybridization and dynamic genomic changes. This chromosomal reshuffling is poorly understood for most polyploid species, despite their evolutionary and agricultural importance, due to the challenge of distinguishing homologous sequences from each other. Here, we use dense linkage maps generated with targeted sequence capture to improve the diploid strawberry (*Fragaria vesca*) reference genome and to disentangle the subgenomes of the wild octoploid progenitors of cultivated strawberry,*Fragaria virginiana* and *Fragaria chiloensis*. Our novel approach, POLiMAPS (Phylogenetics of Linkage-Map-Anchored Polyploid Subgenomes), leverages sequence reads to associate informative interhomeolog phylogenetic markers with linkage groups and reference genome positions. In contrast to a widely accepted model, we find that one of the four subgenomes originates with the diploid cytoplasm donor*F. vesca,* one with the diploid *Fragaria iinumae,* and two with an unknown ancestor close to *F. iinumae*. Extensive unidirectional introgression has converted *F. iinumae*-like subgenomes to be more *F. vesca*-like, but never the reverse, due either to homoploid hybridization in the *F. iinumae*-like diploid ancestors or else strong selection spreading *F. vesca*-like sequence among subgenomes through homeologous exchange. In addition, divergence between homeologous chromosomes has been substantially augmented by interchromosomal rearrangements. Our phylogenetic approach reveals novel aspects of the complicated web of genetic exchanges that occur during polyploid evolution and suggests a path forward for unraveling other agriculturally and ecologically important polyploid genomes.

INTRODUCTION

Whole-genome duplication has occurred frequently in the evolution of flowering plants as well as other taxa, which has resulted in genomes composed of multiple homeologous subgenomes (Otto and Whitton 2000;Wood et al. 2009; Jiao et al. 2011). Polyploidy can evolve through interspecific hybridization (allopolyploidy) or within a single lineage (autopolyploidy), and higher-order polyploids (>4*x*) may have a complex history of multiple allo- and/or autopolyploid duplications in addition to homoploid hybridization events that do not change chromosome number (Marcussen et al. 2014). Following duplication, evolutionarily successful polyploids rapidly undergo radical genomic changes including gene loss, gene conversion, transposition, and increasing diploidization (Fontdevila 2005; Madlung et al. 2005;

Chen and Ni 2006; Woodhouse 2010; Feldman and Levy 2012). These dynamic changes compensate for the biochemical "genomic shock" of suddenly having multiple, perhaps divergent copies of all genes, while retaining the fitness benefits of a large, diverse, and versatile genome (Chen and Ni 2006; Hollister et al. 2012). Introgression of DNA from one allo-subgenome to another can be due to homoploid hybridization between the diploid ancestors prior to polyploidization (Marcussen et al. 2014), or it can occur postpolyploidization through homeologous exchange that may be initially reciprocal (crossing over) or nonreciprocal (gene conversion) (Kovarik et al. 2004; Wang et al. 2009;Gaeta and Pires 2010; Salmon et al. 2010; Chalhoub et al. 2014). Transposition mediated by transposable elements occurs often in plants and other organisms (Lisch 2013), so polyploid subgenomes will likely show variation in copy number at loci that duplicated either in an ancestor prior to allopolyploidization or else in the polyploid as a direct response to whole-genome duplication (Chen and Ni 2006). Polyploidy plays a central role in speciation and adaptation, but the evolutionary relationships and genomic interactions among subgenomes are rarely well understood outside of a few heavily studied agriculturally important species (e.g., Xiong and Pires 2011; Feldman and Levy 2012; Page et al. 2013; Marcussen et al. 2014), due to the technical challenge of separating homeologous sequence. Phylogenetic relationships of homeologous chromosomes have been estimated from cytological segregation patterns and FISH (fluorescence in situ hybridization) karyotyping (e.g., Bringhurst 1990; Maluszynska and Hasterok 2005; Lipman et al. 2013), or from individual gene sequences (e.g., Senchina et al. 2003; Shimizu-Inatsugi et al. 2009; Rousseau-Gueutin et al. 2009; Cenci et al. 2012; DiMeglio et al. 2014). However, a full picture of the duplication and subsequent evolution of genomes requires genome-scale sequence data sets with representative markers spanning entire chromosomes.

The strawberries (*Fragaria*) present an excellent system for studying the evolution of polyploidy (Liston et al. 2014). Within a short time (1–4 Ma), a diploid ($2n = 2x = 14$ chromosomes) ancestral *Fragaria* has diversified into 20 species, nearly half of which are polyploid (Njuguna et al. 2013). Two wild octoploids, *Fragaria virginiana* and *Fragaria chiloensis*, are sister species and the progenitors of the cultivated octoploid strawberry *F. ×ananassa* (Njuguna et al. 2013). Phylogenetics based on whole chloroplast genome sequencing places these two species and the octoploid/decaploid*Fragaria iturupensis* as sister to the North American diploid *Fragaria vesca*ssp. *bracteata,* suggesting that the latter is the cytoplasmic donor to the higher-order polyploids (Njuguna et al. 2013). Phylogenetic analysis of nuclear genes indicates that the octoploids have an allopolyploid history with ancestry in both the *F. vesca* clade and a clade containing the Japanese diploid *Fragaria iinumae* (Rousseau-Gueutin et al. 2009). The octoploid subgenomes are highly diploidized and inheritance is thought to be primarily disomic (Byrne and Jelenkovic 1976; Bringhurst 1990;Ashley et al. 2003; Rousseau-Gueutin et al. 2008), although there is some evidence for a small amount of polysomic inheritance (Lerceteau-Köhler et al. 2003). It remains unclear whether subgenomes largely act as independent evolutionary units or whether polysomic recombination over many generations blurs this distinction. Multivalent pairings observed in interspecific hybrids have been used to infer relationships among subgenomes. The most widely accepted cytological octoploid genome formula is AAA′A′BBB′B′ based on segregation patterns (Bringhurst 1990;Sargent et al. 2012) and phylogenetic analysis of individual genes (Rousseau-Gueutin et al. 2009; DiMeglio et al. 2014). However, numerous cytological studies over the past century have suggested various contradictory models (Ichijima 1926; Yarnell 1931; East 1934; Federova 1946; Senanayake and Bringhurst 1967; Nathewet et al. 2010). Although genome-scale methods have been applied to octoploid *Fragaria* (e.g.,Isobe et al. 2013; Hirakawa et al. 2014; van Dijk et al. 2014), these studies did not isolate phylogenetic markers differentiating the subgenomes. Thus, the relationships among homeologous chromosomes remain unclear. The size of the octoploid *Fragaria* genome is approximately 80% of a strict quadrupling of the diploid genomes (Hirakawa et al. 2014), suggesting that substantial gene loss has occurred postpolyploidization. Genetic studies of polyploid *Fragaria* will be facilitated by clear demarcations of the subgenomes, essential for both basic and applied research in this evolutionarily and agriculturally important genus (Liston et al. 2014).

Here, we leverage several key resources and strategies, including an improved reference genome, dense linkage maps from crosses of highly heterozygous parents, and high-throughput next-generation sequence data, to dissect the subgenomes of octoploid *Fragaria*. We present a novel approach, POLiMAPS (Phylogenetics Of Linkage-Map-Anchored Polyploid Subgenomes; fig. 1), which allows resolution of previously intractable basic questions about polyploid genome organization and is directly applicable to many other important taxa. Our results revise and enhance the accepted model of evolutionary relationships among subgenomes, and highlight the complex history of genetic exchanges that can underlie polyploid genomes.

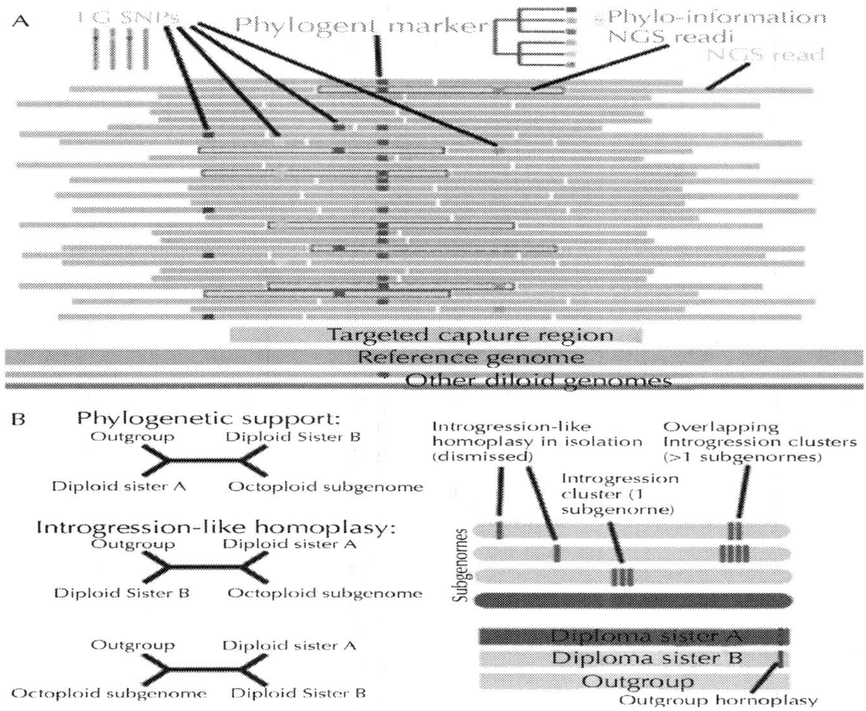

Figure 1: Schematic of our POLiMAPS approach. (*A*) Next-generation sequence reads from octoploids obtained through targeted capture are aligned to our reference genome assembly. LG SNPs occur at approximately one-eighth frequency in a single parent and show Mendelian segregation in the offspring. If an LG SNP shares a read with a marker that occurs on multiple linkage groups and/or diploid taxa, this marker can be used in phylogenetics. A majority of reads do not harbor both an LG SNP and a phylogenetic marker, and thus many linkage groups (e.g., dark blue) have missing data for many phylogenetic markers. (*B*) Identification of introgression. Phylogenetic markers can be classified as providing phylogenetic support, showing introgression-like homoplasy, or showing outgroup homoplasy. Multiple adjacent markers showing introgression-like homoplasy, interspersed with few or no phylogenetically supportive markers, are considered to be caused by true introgression.

MATERIALS AND METHODS

New Fvb Genome

We revised the assembly of the *F. vesca* reference genome (v. 1.1, here designated FvH4; Shulaev et al. 2011) based on linkage map data from three parental plants of *F. vesca* ssp. *bracteata*. Our justification for this new assembly is 2-fold. First, the high density of segregating polymorphisms in our linkage maps relative to the *F. vesca* ssp. *vesca* × *Fragaria bucharica*linkage map used in the original genome assembly (Sargent et al. 2011) provides us with greater accuracy in ordering and orienting scaffolds. Second, for any true biological differences in genome structure between *F. vesca* ssp. *bracteata* and either *F. vesca* ssp. *vesca* or *F. bucharica*, the *F. vesca* ssp. *bracteata* arrangement is preferred because this subspecies is the cytoplasmic donor to the octoploid clade (Njuguna et al. 2013). We previously described linkage maps from an outbred cross between two individuals from a single population in Oregon (Tennessen et al. 2013), referred to here as Fvb-m and Fvb-p ("maternal" female ovule donor, "paternal" hermaphrodite pollen donor). Here, we additionally generated a linkage map from F1 offspring derived from selfing a hermaphroditic *F. vesca* ssp. *bracteata* plant collected from Lincoln National Forest, Cloudcroft, NM (32.968 N, −105.750 W) (supplementary table S1, (hereafter Fvb-s for "selfed"—both ovule and pollen donor). Following our previously described methodology (Tennessen et al. 2013), we extracted DNA from 43 F1 offspring and target captured sequences using our published probes (Tennessen et al. 2013). Our probes were originally designed as sets of three partially overlapping 100-bp oligonucleotides surrounding 6,575 central targeted polymorphisms. The targeted polymorphisms were not necessarily segregating in any of the plants examined in this study, but we expected that the probes would often capture other single nucleotide polymorphisms (SNPs). Samples were sequenced in multiplex on the Illumina HiSeq 2000 at Oregon State University. We retained 41 offspring with high sequencing coverage for analysis. Sequencing data in FASTQ format are uploaded to the NCBI SRA (Bioproject Accession PRJNA263688). As in Tennessen et al. (2013), we used BWA (Li and Durbin 2009) to map our reads to a supplemented reference genome that included scaffolds left out of FvH4 (v. 1.1) and converted genotypes to vcf format using SAMtools (Li et al. 2009). We retained genotypes as valid only if per-individual depth was at least 20, the Phred-scaled likelihood was 0, and Phred-scaled likelihood for all other potential genotypes was ≥40 (likelihood of other genotype $\leq 10^{-4}$); otherwise, genotypes were considered missing. We retained putative polymorphisms for linkage mapping if they showed missing genotypes in fewer than eight offspring, did not show greater than 85% of offspring with the same genotype, and had informative parental genotypes consistent with segregation in the offspring.

We used OneMap (Margarido et al. 2007) to generate a linkage map for Fvb-s. Initially, we only used polymorphic sites with no missing genotypes to create the framework linkage map; sites with missing data were subsequently added manually if possible. We used a logarithm of odds (LOD) of 5 to assign segregating polymorphisms to linkage groups, and polymorphisms were assigned to the most likely position within the linkage group based on their LOD score. Genotyping and mapping errors were identified and corrected manually as described in Tennessen et al. (2013).

We compared our Fvb-s map with our previous Fvb-m and Fvb-p maps, and tested whether the position of any polymorphism was inconsistent among maps. We then used these maps to generate a new genome assembly that placed all polymorphisms in linear order based on their linkage map positions. To do so, we assumed that scaffolds were assembled correctly, and therefore we rearranged the order of whole scaffolds rather than breaking scaffolds into sections, unless we had multiple polymorphisms on the same scaffold mapping to different locations. We defaulted to the scaffold order and orientation in FvH4 unless we had conflicting evidence from our linkage maps. We used the*Prunus persica* (peach) genome (International Peach Genome Initiative et al. 2013) to guide two particular types of decision: The specific locations of scaffold splits and the placement of interchromosome translocations that could not be placed unambiguously based on map position. Thus, we only used the *Prunus* genome as a guide when we already needed to alter the FvH4 assembly based on our linkage maps, and never used the *Prunus*genome alone to override the FvH4 assembly. In order to assess whether FvH4 or Fvb more accurately reflects gene order, we compared synteny of both assemblies with that of the *Prunus* genome. We first used BLAT (Kent 2002) to identify putative orthologs. We then examined the order of*Prunus* orthologs in the two *Fragaria* genomes and vice versa. For each assembly, we counted stretches of continuously syntenous genes (pertaining to the same chromosome, even if their order is rearranged).

Octoploid Maps

In order to reconstruct a detailed evolutionary history of octoploid strawberry subgenomes, we examined two linkage maps from crosses that have previously been described and initially mapped with simple sequence repeat (SSR) markers: One for *F. virginiana* ssp. *virginiana*(collected from Pennsylvania; Spigler et al. 2008, 2010, 2011) and one for*F. chiloensis* (collected from Oregon; Goldberg et al. 2010). Both crosses consist of the F1 offspring of two unrelated outbred parents (a female ovule donor and a hermaphrodite pollen donor in each cross). We employed targeted capture sequencing which increased the density of segregating polymorphisms by approximately an order of magnitude, provided each linkage group marker (LG SNP) with a clear expected position in the reference genome, and allowed us to identify phylogenetic markers corresponding to specific linkage groups. DNA was extracted from 100 mg fresh young leaf tissue from F1 offspring (73 for *F. virginiana* ssp. *virginiana* and 46 for *F. chiloensis*) using a modified CTAB (cetyltrimethylammonium bromide) procedure (Doyle JJ and Doyle JL 1987). Targeted capture and sequencing were performed as described above and in Tennessen et al. (2013) (Bioproject Accession PRJNA263688). We analyzed 67 *F. virginiana* F1 offspring and 42 *F. chiloensis* F1 offspring with high sequencing coverage. In addition, we sequenced the complete genomes of the maternal parent in both octoploid crosses, following our previous methodology for low coverage whole-genome sequencing of *Fragaria* (Tennessen et al. 2013).

For both octoploid crosses, we mapped reads to the Fvb genome assembly using BWA. Because SAMtools cannot call variants in polyploids, we generated a pileup format file for each cross and used a custom Perl script to call polymorphisms (Perl script available athttps://github.com/listonlab/POLiMAPS). Our goal was to identify variants occurring with approximately one-eighth frequency (i.e., heterozygous at a single homeolog), which segregated in a Mendelian fashion (e.g., if heterozygous in one parent then heterozygous in approximately half of the progeny). Fixed differences between homeologs can be distinguished from true SNPs because they will not show Mendelian segregation. We coded genotypes as heterozygous if the rare variant occurred at least twice and at ≥2.5% frequency. We coded genotypes as homozygous if there was no polymorphism or if the rare variant occurred a single time and at less than 1.25% frequency, which we considered to be a potential sequencing error. Ambiguous genotypes that did not meet these criteria or which had less than 32× coverage were coded as missing. To filter for Mendelian segregation, we only retained sites in which the two genotypes were each observed in at least eight progeny. We retained markers for linkage mapping (LG SNPs) only if they had missing data for no more than a single progeny. We converted remaining genotypes to OneMap format and used OneMap to generate linkage maps as with Fvb-s. Due to the complexity of octoploid linkage groups, we did not attempt to manually add LG SNPs with larger numbers of missing genotypes, as we did with the diploid Fvb-s. For both octoploid species, we used a LOD of 5 to assign LG SNPs to linkage groups, and LG SNPs were assigned to the most likely position within the linkage group based on their LOD score. Some linkage groups were then manually joined based on visual inspection of segregation patterns, Fvb position and phylogenetic position. "Minor" linkage groups with fewer than five LG SNPs are reported but excluded from subsequent analysis. For all linkage groups, we assigned a haploid chromosome number (1–7) corresponding to the Fvb pseudochromosome with the largest number of LG SNPs in the linkage group. Each "major" linkage group of ≥5 LG SNPs was then named for the species, haploid chromosome number (Roman numeral matching previous map designations; Goldberg et al. 2010; Spigler et al. 2010;, subgenome (determined phylogenetically; see below), and parent (p for paternal or m for maternal); for example, "Fvirg-IV-Av-p" is a linkage group from *F. virginiana,* with most of its LG SNPs coming from Fvb4 (and previously designated IV by Spigler et al. 2010), with phylogenetic subgenome designation Av, and with paternally segregating LG SNPs. For most analyses, we ignored whether SNPs fell within the targeted region or in additional nontargeted captured sequence. Adjacent SNPs less than 1 kb apart (approximately the width of adjacent high-coverage sequence captured by a set of overlapping probes) were considered to be distinct LG SNPs but in the same "region," whether or not that region was targeted with our probes. For some analyses (described below), we restricted the data to only targeted regions.

Phylogenetic Analysis

For all LG SNPs in the octoploid linkage groups, we identified all original octoploid Illumina reads that mapped to Fvb and contained the minor variant (fig. 1*A*). We used these reads to infer the sequence of the subgenome in the vicinity of the LG SNP (Perl script available athttps://github.com/listonlab/POLiMAPS). In cases where a linkage group had an LG SNP less than a read length away from a variant differentiating other linkage groups and/or diploid taxa, we used this adjacent variant as a phylogenetic marker. We did not use the LG SNPs themselves in phylogenetic analysis because they are variable within the linkage group and they were chosen such that the minor variant only occurred in a single subgenome, so we did not expect any meaningful phylogenetic signal from the LG SNPs.

We sequenced whole genomes of *F. iinumae*, *Fragaria mandshurica*, and*Fragaria viridis* for use in phylogenetic analysis, following our previous methodology for low coverage whole-genome sequencing of *Fragaria* (Tennessen et al. 2013) (Bioproject Accession PRJNA263688). Using BWA, we mapped reads from these three species, as well as our original Fvb-s, Fvb-m, and Fvb-p reads, to Fvb and generated a vcf file with SAMtools. In addition, we included the published contigs of *Fragaria nipponica*(http://www.ncbi.nlm.nih.gov/assembly/GCA_000512025) and *F. bucharica* (http://www.ncbi.nlm.nih.gov/assembly/GCA_000511995) and used BLAT with default parameters to match contigs to Fvb (USDA accession CFRA522 is *F. bucharica* but was identified as *Fragaria nubicola*in Hirakawa et al. 2014). Because the octoploid subgenomes originated from a *F. vesca*-like ancestor and a *F. iinumae*-like ancestor, we refer to these two species along with the octoploid subgenomes as "ingroup" taxa, whereas all diploid species other than *F. vesca* and *F. iinumae* are "outgroup" species, even though the ingroup taxa do not form a monophyletic group with each other. As an outgroup to the entire genus, we chose the transcriptome of *Rubus coreanus*(http://www.ncbi.nlm.nih.gov/sra/SRX347804; Hyun et al. 2014). Because*Rubus* is too distant from *Fragaria* to reliably map reads with BWA, we used BLAT with default parameters to match reads to Fvb. We attempted to use *Potentilla micrantha* as an additional outgroup based on nuclear reads from the sequencing of its chloroplast genome (Ferrarini et al. 2013) but coverage was too low to be useful. Using RAxML (Stamatakis 2006) with -N autoMRE and -m GTRCAT, we estimated separate phylogenies for each of the seven haploid *Fragaria* chromosomes from each of the four parents of the octoploid maps and the three parents of the diploid maps (matrices and trees available:http://purl.org/phylo/treebase/phylows/study/TB2:S15849).

Introgression

We tested for evidence of introgression between *F. vesca*-like and *F. iinumae*-like subgenomes, as can occur through homoploid hybridization before polyploidization or homeologous exchange after polyploidization. Because introgression will manifest as inconsistencies in the phylogenetic signal, we first identified sites showing homoplasy. For a site with a known base for at least one octoploid subgenome, *F. vesca*, *F. iinumae*, and at least one outgroup species, there are three possible phylogenetic patterns (fig. 1*B*). First, the site would be supportive of the phylogeny if the subgenome shares a base with its sister ingroup species (either *F. vesca* or*F. iinumae*), whereas the other two species share a different base. Second, the site would show "introgression-like homoplasy" if the subgenome shares a base with the wrong ingroup species, whereas its sister species shares a different base with the outgroup, consistent with introgression of DNA between the two ingroup lineages. Third, the site would show "outgroup homoplasy" if the subgenome shares a base with the outgroup, whereas *F. vesca* and *F. iinumae* share a different base. In the absence of true introgression, homoplasy is only due to factors such as independent mutations, incomplete lineage sorting, and sequencing errors, and thus introgression-like homoplasy should occur at the same prevalence as outgroup homoplasy. Therefore, we tested for an excess of introgression-like homoplasy in order to detect introgression (similar to the ABBA/BABA test; Green et al. 2010).

For probes with introgression-like homoplasy, we took two additional steps to distinguish between true introgression and spurious homoplasy due to other evolutionary processes such as repeated independent mutations. First, probes with both supportive and introgression-like homoplasy sites for the same parental sample were classified as "support/homoplasy" probes and were considered to not show true introgression. Second, we identified clusters of at least two adjacent introgression-like homoplasy probes in the Fvb genome, such that the ratio of introgression homoplasy probes to supportive probes in the region was at least 4. Only these clusters were considered to be true introgression events. In order to assess whether introgression occurred before or after polyploidization, we tested whether introgressed chromosomes showed a closer phylogenetic relationship with their homeologous chromosomes or with the sister ingroup diploid species.

As an independent test of introgression, we examined the relative depth of *F. vesca*-like and *F. iinumae*-like alleles across the low-coverage genomes of the maternal parents of both octoploid crosses. We identified SNPs distinguishing *F. vesca* and *F. iinumae* that were also polymorphic in the octoploids, with a combined depth across the two maternal parents between 45× and 200×. We calculated the depth ratio for both variants, excluding sites with a ratio greater than 8 or less than 1/8 as likely errors, and tested whether this ratio differed between introgression clusters and the rest of the genome.

To look for correlations between overall depth and introgression, we measured depth at the central targeted site of each targeted region in the Fvb-s parent and the octoploid parents. We normalized depth for each individual by dividing by the sum of depths at all central targeted sites, and for each targeted region we calculated the mean normalized depth across octoploids. We then calculated the ratio of normalized depth in octoploids versus Fvb-s, and we tested whether this ratio differs among regions with or without introgression.

Interchromosomal Rearrangements

In order to identify interchromosome rearrangements, we first incorporated data from a recent linkage map of *F.* × *ananassa* (Isobe et al. 2013) along with our octoploid linkage maps. We used the primer sequences for each SSR marker (Isobe et al. 2013) and we performed in silico polymerase chain reaction (PCR) with BLAT (Kent 2002) on our Fvb assembly. We retained markers for which both forward and reverse primers mapped within 2 kb of each other. To assess translocations among subgenomes and species, we assigned all linkage groups in *F.* ×*ananassa, F. virginiana* ssp. *virginiana*, and *F. chiloensis* to a haploid chromosome that matched the majority of its markers (SSR or SNPs), and then we identified markers with an Fvb position that did not match this primary chromosome. We tested whether any translocations matched FvH4 but not Fvb, which could indicate errors in our Fvb assembly. We tested whether rearrangements were significantly clustered in their reference genome position by generating 10,000 simulated data sets in which LG SNPs showing rearrangements were randomly distributed among all targeted regions with LG SNPs. As with introgression (above), we tested whether the ratio of octoploid:diploid normalized depth differed between targeted regions with or without interchromosomal rearrangements.

RESULTS

New Fragaria Reference Genome

In the targeted capture sequence data from the 41 Fvb-s progeny, 3.244 Mb had a mean per-individual depth ≥20×. Within these high-coverage sites, 1,825 polymorphic sites passed our quality thresholds and were used in linkage mapping. Our initial Fvb-s linkage map consisted of eight linkage groups corresponding to the seven *Fragaria* chromosomes with two linkage groups for different sections of FvH4 pseudochromosome 4, which were manually joined for a final Fvb-s linkage map of seven linkage groups spanning 326 cM.

We found no discrepancies among the Fvb-m, Fvb-p, and Fvb-s linkage maps. In contrast, between FvH4 and our *F. vesca* ssp. *bracteata* maps, we observed 44 interchromosome translocations, 40 intrachromosome translocations, 39 inversions, and 18 placements of unmapped FvH4_0 scaffolds (87 of these 141 rearrangements are the same as those previously described, Tennessen et al. 2013). We made 39 splits among the 246 FvH4 scaffolds and added two missing scaffolds (scA and scB as described in Tennessen et al. 2013) for a total of 287 new scaffolds, which we rearranged to match the *F. vesca* ssp. *bracteata* maps. Our new assembly, here designated Fvb, contains 208.9 Mb of scaffold sequence, with 207.0 Mb assembled into seven pseudochromosomes (Fvb1–Fvb7) and 1.9 Mb remaining unassembled on a false chromosome Fvb0. With 10-kb gaps separating all scaffolds, the complete size of Fvb is 211.7 Mb (Available on Figshare: http://dx.doi.org/10.6084/m9.figshare.1259206). We observed 164 distinct stretches of contiguous *Prunus* orthologs in FvH4, but only 116 such stretches in Fvb (fig. 3).

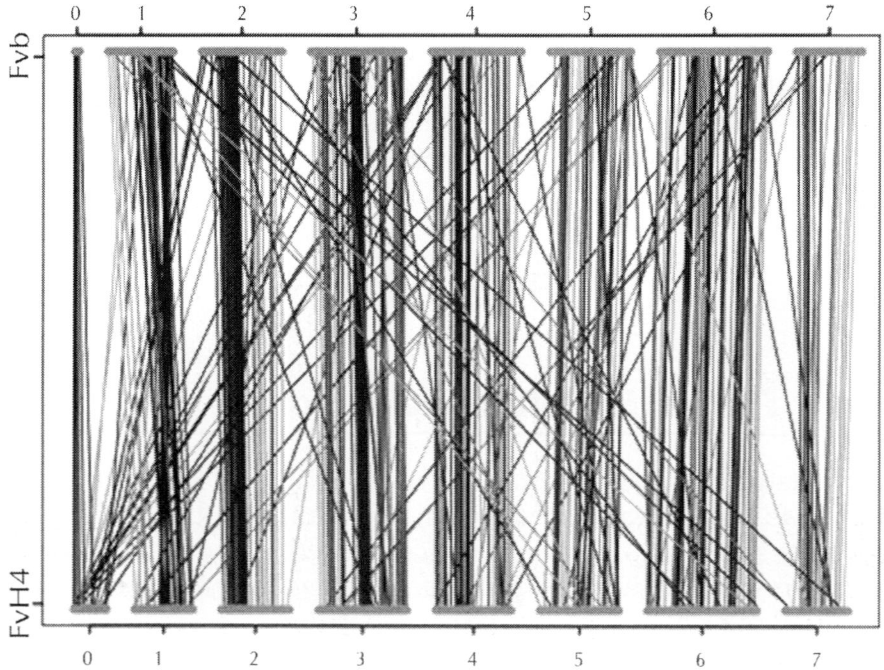

Figure 2: Mapping of scaffolds from FvH4 reference genome assembly to Fvb assembly. The path of every scaffold or scaffold segment is represented by a line (orange, inverted in diploid linkage maps; purple, not inverted in diploid linkage maps; black, no information about scaffold orientation in diploid linkage maps, retained default noninverted orientation).

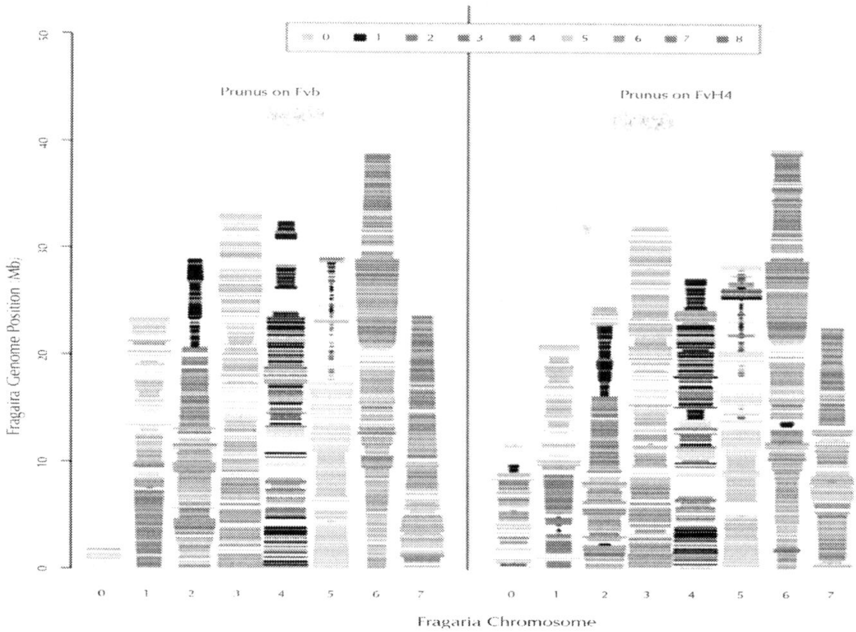

Figure 3: *Prunus* orthologs mapped onto *Fragaria* assemblies. Each horizontal line represents an orthologous gene, colored according to its *Prunus* chromosome, and with a width corresponding to its position on this *Prunus*chromosome (wider lines are close to the start of their respective *Prunus*chromosome). Its *x* axis position indicates the*Fragaria* pseudochromosome onto which it is assembled. Its *y*axis position indicates its position on this *Fragaria*pseudochromosome. Thus, regions showing lines of the same color and similar width indicate high synteny between the two genera. Overall, Fvb shows higher synteny than FvH4.

Octoploid Maps

In the targeted capture reads from the two *F. virginiana* ssp. *virginiana*parents and their 67 progeny, 1,506 kb had a mean per-individual depth ≥32×, whereas 553 kb had a mean per-individual depth ≥80×. A total of 3,875 LG SNPs passed our quality thresholds and were used in linkage mapping, including 1,899 maternal LG SNPs and 1,976 paternal LG SNPs. The additional 594 LG SNPs polymorphic in both parents were not placed in the linkage maps but were used to identify pairs of corresponding linkage groups between the parents.

Using a LOD threshold of 5 initially yielded 27 maternal linkage groups and 29 paternal linkage groups. We split all linkage groups anywhere there was a gap greater than 33 cM, resulting in a 34 maternal linkage groups and 35 paternal linkage groups. Some of these linkage groups were then manually joined based on visual inspection of segregation patterns, position of these groups of LG SNPs in the Fvb assembly (e.g., if they corresponded to the same Fvb pseudochromosome), and phylogenetic position (see below). Our final count was thus 28 major linkage groups (≥ 33 LG SNPs each) for both parents, representing the 28 haploid chromosomes, plus 4 maternal and 5 paternal minor linkage groups (≤ 3 LG SNPs each) (fig. 4).

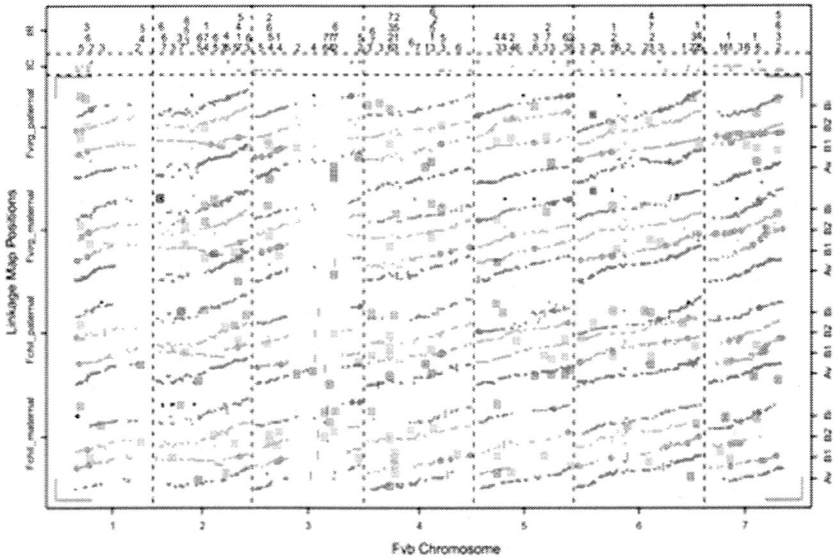

Figure 4: Linkage maps for *F. chiloensis* and *F. virginiana*ssp. *virginiana*, aligned to the Fvb reference genome. Segregating LG SNPs are represented by small solid circles. For all LG SNPs, the *y*axis position corresponds to linkage map position (cM) and the *x* axis position corresponds to Fvb genome position (Mb). All seven Fvb chromosomes are represented on the *x* axis, with scale indicated by dark gray horizontal bars in all four corners (=10 Mb). All four parents are represented on the *y* axis, with scale indicated by light gray vertical bars in all four corners (=100 cM). Linkage groups are colored based on their phylogenetic position (red, Av; cornflower, B1; cyan, B2; blue, Bi). LG SNPs in black correspond to minor linkage groups (<5 LG SNPs) that could not be placed in the phylogeny. Short vertical black bars indicate junctures where

smaller linkage groups were manually joined based on visual inspection of genotypes, phylogenetic position, and Fvb position. Introgression clusters are shown as light red circles behind LG SNPs, and their positions are again shown along the top of the figure (IC), colored according to subgenome. Interchromosome rearrangements are shown as squares around LG SNPs, and their positions are again shown along the top of the figure (IR), with numbers indicating the haploid chromosome of the linkage group.

In the targeted capture reads from the two *F. chiloensis* parents and their 42 progeny, 1,632 kb had a mean per-individual depth ≥32×, whereas 705 kb had a mean per-individual depth ≥80×. A total of 2,542 LG SNPs passed our quality thresholds and were used in linkage mapping, including 1,100 maternal LG SNPs and 1,442 paternal LG SNPs. As with *F. virginiana* ssp. *virginiana*, an additional 780 LG SNPs polymorphic in both parents were used to pair corresponding linkage groups. Using a LOD threshold of 5 initially yielded 39 maternal linkage groups and 36 paternal linkage groups. Splitting at gaps greater than 33 cM resulted in 39 maternal linkage groups and 40 paternal linkage groups. As with *F. virginiana* ssp. *virginiana*, some of these linkage groups were then manually joined, resulting in 28 major linkage groups (≥17 LG SNPs each) for both parents, representing the 28 haploid chromosomes, plus 2 maternal and 2 paternal minor linkage groups (≤ 4 LG SNPs

each) (fig. 4).

Phylogenetic Analysis

We obtained 540–1,517 markers per haploid chromosome for phylogenetic analysis, 317–902 of which were phylogenetically informative with respect to the relationship between an octoploid subgenome, *F. vesca*, *F. iinumae*, and one of the diploid outgroup species These markers represented 160–421 regions of 1 kb in Fvb (table 1). For any given site, we could only include a linkage group if it had an LG SNP less than read length away (fig. 1*A*), so at most sites the majority of taxa had missing data (64% of total data missing). Nevertheless, phylogenetic signal was sufficient to resolve trees with high bootstrap support (usually greater than 90%) for the major clades at all seven chromosomes (fig. 5; matrices and trees available:http://purl.org/phylo/treebase/phylows/study/TB2:S15849). In all seven trees, one of the four octoploid subgenomes groups with *F. vesca*, which we designated subgenome Av. Chromosomes from the other three subgenomes consistently form a clade with *F. iinumae*. The subgenome sister to *F. iinumae* we designated subgenome Bi, and the two basal to the*F. iinumae*/Bi clade we designated subgenomes B1 and B2, with B1 being the subgenome showing greatest divergence from *F. iinumae*. Because this divergence criterion is weak relative to the topology-based criteria, this distinction may not be evolutionarily meaningful across haploid chromosomes of subgenomes B1 versus B2.

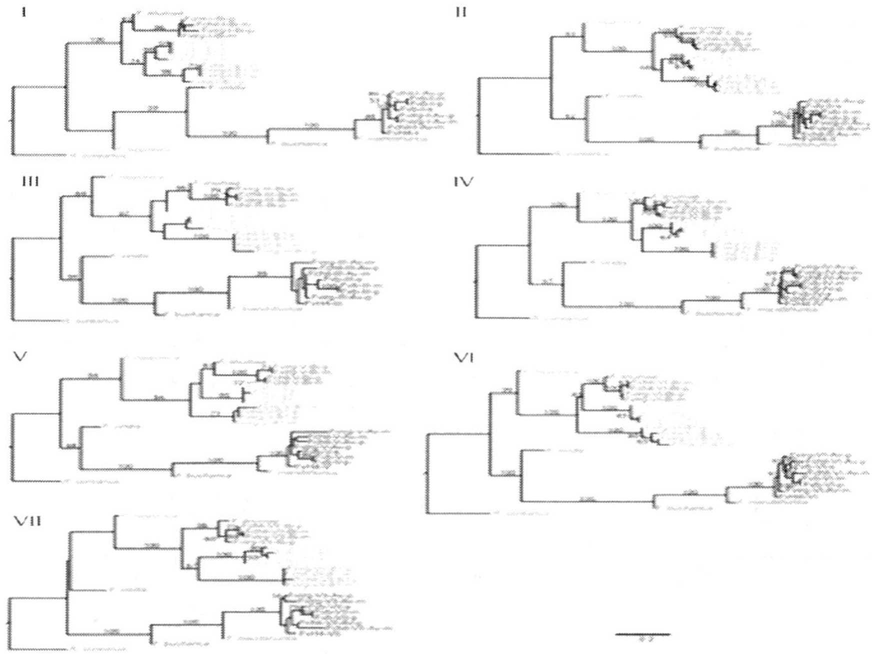

Figure 5: Phylogenies of all seven haploid *Fragaria*chromosomes. Octoploid subgenomes are named for the species (Fchil or Fvirg), haploid chromosome number (Roman numeral), subgenome (Av, B1, B2, or Bi), and parent (p or m), and are colored as in figure 4with *F. iinumae* (blue) and *F. vesca* (FvH4, Fvb-m, Fvb-p, and Fvb-s; red) similarly colored. Other diploid taxa have distinct colors. Numbers on branches represent bootstrap values; bootstraps less than 50 or on branches with length less than 0.01 substitutions/informative site are not shown.

Table 1: Regions of 1-kb Showing Phylogenetic Support and Homoplasy

Phylogenetic Signal	Subgenome	I LGs	II LGs	III LGs	IV LGs	V LGs	VI LGs	VII LGs	Total
Supportivea									
	Av	54	120	55	97	118	198	90	732
	Bi	44	117	78	93	91	175	102	700
	B1	60	104	43	67	84	120	72	550
	B2	61	104	46	91	86	113	59	560
Clustered introgression -like (clusters)b									
	Av	0 (0)	0 (0)	0 (0)	0 (0)	0 (0)	0 (0)	0 (0)	0 (0)
	Bi	2 (1)	4 (2)	4 (2)	0 (0)	11 (4)	2 (1)	5 (1)	28 (11)
	B1	10 (2)	7 (2)	16 (5)	15 (3)	2 (1)	19 (8)	25 (5)	94 (26)
	B2	6 (3)	0 (0)	0 (0)	0 (0)	4 (2)	3 (1)	17 (5)	30 (11)
Isolated introgression-likec									
	Av	2	4	3	6	4	8	7	34
	Bi	6	4	7	6	11	8	3	45
	B1	4	7	5	6	3	17	12	54
	B2	5	5	7	9	5	10	9	50
Outgroup homoplasyd									
	Av	2	6	3	6	6	10	18	51
	Bi	1	2	4	3	3	10	7	30
	B1	2	7	5	5	3	14	16	52
	B2	8	6	5	2	10	19	13	63
Support/ homoplasye									
	Av	1	5	3	6	5	7	4	31
	Bi	1	6	1	3	4	8	8	31
	B1	3	9	3	8	6	16	10	55
	B2	2	10	6	5	10	11	13	57
Total informative regionsf	All	160	298	185	230	255	421	264	1,813
Total informative markersg	All	317	645	367	483	568	902	536	3,818

Total markers[h]	All	540	1, 152	627	784	905	1, 517	956	6,481
Total missing[i]	All	64.5	64.5	64.5	63.2	64.1	63.9	64.6	64.2

[a]Regions which support the phylogenetic position of a given subgenome.

[b]Regions showing homoplasy such that a subgenome's relationships to *Fragaria iinumae* and *Fragaria vesca* are switched, which occur in clusters near other such regions. We conclude that these are true introgression events.

[c]Regions showing homoplasy such that a subgenome's relationships to *F. iinumae* and *F. vesca* are switched, which do not occur in clusters. These may be additional introgression events but we do not analyze them as such.

[d]Regions showing homoplasy such that a subgenome matches an outgroup taxon and not *F. iinumae* and *F. vesca.*

[e]Regions with sites that both support the position of a parental sample and show homoplasy for the same parental sample.

[f]Regions containing at least one phylogenetic marker that is informative with respect to the relationship between an octoploid subgenome, *F. vesca, F. iinumae,* and a third outgroup taxon (informative markers).

[g]Count of informative markers.

[h]Count of total phylogenetic markers.

[i]Percentage of missing data.

Introgression

We observed phylogenetic support at 54–198 regions per haploid chromosome (732 total) for subgenome Av, 44–178 regions (700 total) for subgenome Bi, 43–120 regions (550 total) for subgenome B1, and 46–113 regions (560 total) for subgenome B2 (table 1; fig. 6). We also observed a large number of regions showing homoplasy with respect to the consensus trees, that is, they are inconsistent with the phylogeny: 116 for Av, 134 for Bi, 255 for B1, and 200 for B2 (table 1; fig. 6). To find introgression homoplasy, we considered regions where subgenome Bi, B1, or B2 is closer to *F. vesca* than to *F. iinumae,* or where subgenome Av is closer to *F. iinumae* than to *F. vesca* (fig. 1*B*). Introgression homoplasy was 1.7 times as common as outgroup homoplasy (335 vs. 196; $P < 0.0001$; Fisher's exact test; table 1). The most common type of homoplasy was when subgenome Bi, B1, or B2 appeared closer to *F. vesca* than to *F. iinumae,* with more than twice as many regions showing this type of homoplasy than would be expected given the outgroup homoplasy (301 vs. 145; $P < 0.0001$;

Fisher's exact test; table 1). Because individual instances of homoplasy could be due to independent mutations or incomplete lineage sorting rather than true introgression, we focused on the 152 regions that formed 48 distinct genomic clusters containing multiple (2–11) regions showing introgression homoplasy (table 2). If these clusters are excluded, there is no longer an excess of introgression homoplasy (fig. 6). These introgression clusters are dispersed throughout the genome, occurring on homeologs of all seven Fvb pseudochromosomes, and encompassing 15.0 Mb (7% of Fvb). Clusters range in size from 2 to 1,579 kb (median = 160 kb; mean = 324 kb). There are only three instances where introgression clusters on different subgenomes overlap, encompassing 604 kb, and introgression regions do not differ from the rest of the genome in relative sequencing coverage (*t*-test, $t = 2.22$, $P > 0.01$; fig. 7*A*).

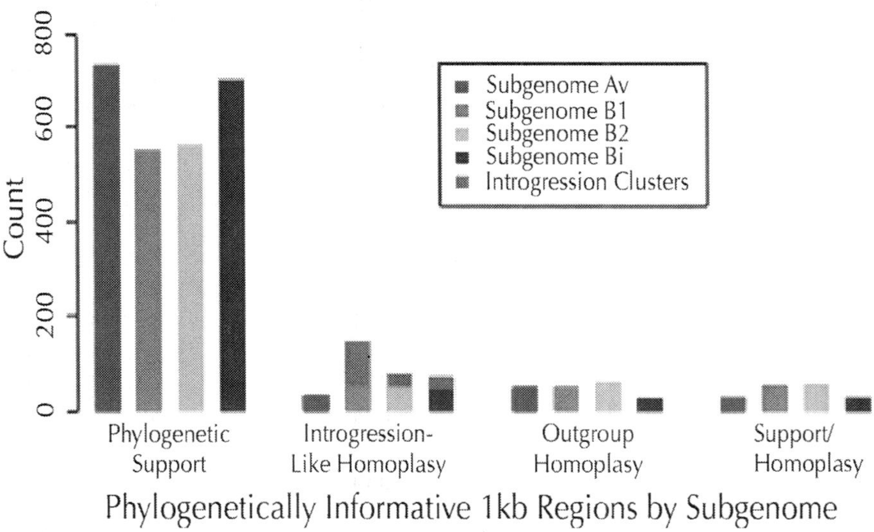

Figure 6: Regions of 1 kb with phylogenetically informative markers. For a phylogeny consisting of an outgroup, *F. vesca*, *F. iinumae*, and a subgenome, there are three possible patterns (fig. 1*B*): Phylogenetic support, introgression-like homoplasy, and outgroup homoplasy. Regions with multiple markers may also show a mix of support and homoplasy. There is an excess of introgression-like homoplasy relative to outgroup homoplasy, but it can be accounted for entirely by the introgression clusters, which occur only on subgenomes B1, B2, and Bi.

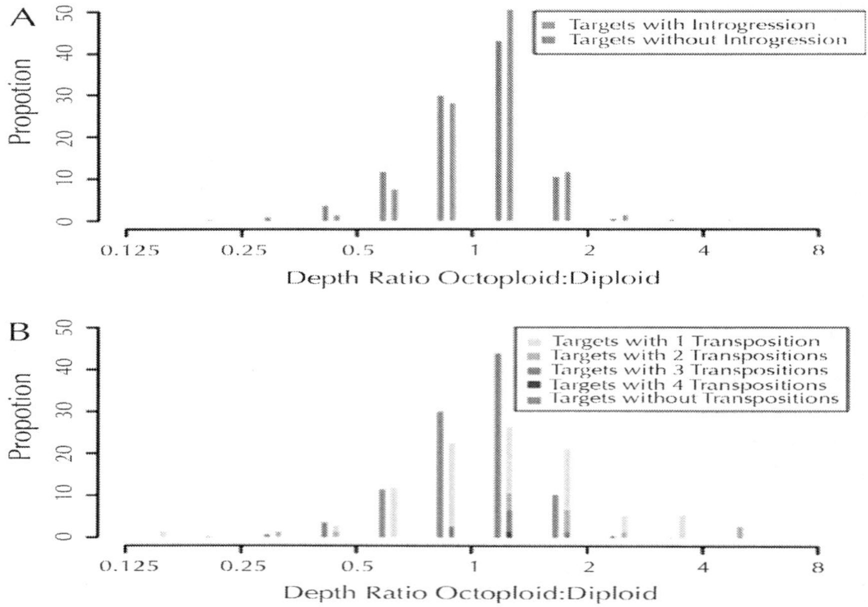

Figure 7: Relative normalized coverage in octoploids relative to diploid, for targeted regions showing dynamic evolutionary changes. Regions are binned on a log scale along the xaxis with breakpoints at multiples of √2. (*A*) Targeted regions showing introgression do not have significantly different depth than regions not showing introgression. (*B*) Targeted regions showing interchromosome rearrangements have higher and more variable depth in octoploid relative to diploid, indicating that many of these regions undergo "copy-and-paste" transposition, resulting in unequal copy numbers across species.

Table 2: Introgression Clusters

Fvb Chromosomea	Fvb Sitesb	Subgenomec	Regionsd	Linkage Mapse
1	5065413–5220128	Bi	2	Fvirg-p
1	427431–447178	B2	2	Fchil-p,Fvirg-p
1	2235284–2394395	B2	2	Fvirg-m,Fvirg-p

1	3893160–4468792	B2	2	Fchil-m,Fvirg-m
1	427431–1390854	B1	8	Fchil-m,Fchil-p,Fvirg-m,Fvirg-p
1	3893160–4468792	B1	2	Fvirg-m,Fvirg-p,Fchil-m
2	22607224–22676892	Bi	2	Fvirg-m,Fvirg-p
2	17223334–17302784	B1	2	Fvirg-m,Fvirg-p
2	25578591–26245090	B1	5	Fchil-m,Fchil-p,Fvirg-m,Fvirg-p
2f	5110972–5139963	Bi	2	Fvirg-m,Fvirg-p
3	30308416–30469404	Bi	2	Fchil-m,Fvirg-p
3	32852305–32955512	Bi	2	Fchil-m,Fchil-p
3	124040–759546	B1	2	Fchil-p,Fvirg-p
3	2067176–2179483	B1	4	Fvirg-m,Fvirg-p
3	4625730–4646210	B1	2	Fvirg-m,Fvirg-p
3	6728933–6815506	B1	3	Fvirg-m,Fvirg-p,Fchil-m
3	7342503–8226577	B1	5	Fchil-m,Fchil-p,Fvirg-m,Fvirg-p
4	22867544–23893840	B1	10	Fchil-m,Fchil-p,Fvirg-m,Fvirg-p
4	26100382–26165868	B1	2	Fchil-m,Fvirg-m
4	32133852–32406162	B1	3	Fchil-m,Fvirg-m
5	275987–481317	Bi	4	Fvirg-m,Fvirg-p,Fchil-p
5	23910116–23947199	Bi	3	Fchil-m,Fchil-p,Fvirg-m
5	28429618–28438057	Bi	2	Fvirg-m,Fvirg-p
5	28912695–29086611	Bi	2	Fvirg-m,Fvirg-p
5	10990292–10992194	B2	2	Fvirg-m,Fvirg-p

5	28429618–28438057	B2	2	Fchil-m,Fchil-p,Fvirg-p
5	17792748–17936726	B1	2	Fvirg-m,Fvirg-p
6	25670424–26071024	Bi	2	Fchil-p,Fvirg-p
6	34884370–34958872	B2	3	Fchil-p
6	127855–704274	B1	4	Fchil-m,Fchil-p,Fvirg-p
6	1024942–1107448	B1	3	Fvirg-m,Fvirg-p,Fchil-p
6	7767792–7780026	B1	2	Fvirg-m,Fvirg-p
6	10737836–10761718	B1	2	Fvirg-m,Fvirg-p,Fchil-m
6	11163923–11407930	B1	2	Fvirg-m,Fvirg-p,Fchil-p
6	27838353–27895788	B1	2	Fchil-m,Fvirg-m
6	34652543–34656025	B1	2	Fchil-m,Fchil-p
6	38080971–38105006	B1	2	Fchil-m,Fvirg-m
7	16085862–16472060	Bi	5	Fchil-m,Fchil-p,Fvirg-m
7	2247584–2857258	B2	2	Fvirg-m,Fvirg-p
7	5329113–5933718	B2	3	Fvirg-m,Fvirg-p
7	6727176–7011896	B2	4	Fchil-m,Fchil-p,Fvirg-m,Fvirg-p
7	8420008–8481933	B2	2	Fchil-m,Fvirg-m
7	9296554–10605847	B2	6	Fchil-m,Fchil-p,Fvirg-m,Fvirg-p
7	7246522–7567127	B1	3	Fchil-m,Fchil-p,Fvirg-p
7	14121260–14312154	B1	2	Fvirg-m,Fvirg-p
7	16665708–18245006	B1	11	Fchil-m,Fchil-p,Fvirg-m,Fvirg-p
7	21644908–22959952	B1	6	Fchil-m,Fchil-p,Fvirg-p

7	23427780–23594132	B1	3	Fvirg-m,Fvirg-p,Fchil-p

[a]Pseudochromosome in Fvb reference genome.

[b]Physical position in Fvb reference genome.

[c]Subgenome designation of linkage group.

[d]Count of adjacent nonoverlapping approximately 1-kb regions contributing to introgression event.

[e]Parental linkage maps in which introgression event is observed.

[f]Primary Fvb chromosome of linkage group is Fvb6 (interchromosome rearrangement).

Strikingly, introgression is unidirectional in that every cluster involves an*F. iinumae*-like subgenome (Bi, B1, or B2) evolving to be more similar to *F. vesca*, with no instances of introgression making subgenome Av more similar to *F. iinumae* (48 vs. 0; compared with the expectation of 36 vs. 12, Fisher's exact test, $P < 0.001$). More specifically, 54% of introgression clusters involve subgenome B1 acquiring *F. vesca*-like sequence. Because we defined subgenome B1 as having greater divergence from *F. iinumae*, introgression may be confounded with the distinction between B1 and B2. However, the average amount of introgression between B1 and B2 (18.5 clusters, 62 regions, 6.9 Mb) is higher than both Bi (11 clusters, 28 regions, 1.7 Mb) and Av (no introgression). In the low-coverage whole-genome data, the mean ratio of *F. vesca*-like reads to *F. iinumae*-like reads is significantly higher within clusters (1.37) than in the rest of the genome (0.98) (*t*-test, $t = 21.96$, $P < 10^{-15}$; fig. 8). Across the genome, there are only 48 sites that can inform whether introgressed regions are closer to *F. vesca* or to the Av subgenome, that is, where a *F. iinumae*-like subgenome has a derived allele with respect to outgroup taxa which is shared by either *F. vesca* or the Av subgenome, but not both. Of these, 23 unite the *F. iinumae*-like subgenome with *F. vesca* to the exclusion of the Av subgenomes, and 25 unite the *F. iinumae*-like subgenome with the Av subgenomes to the exclusion of *F. vesca*.

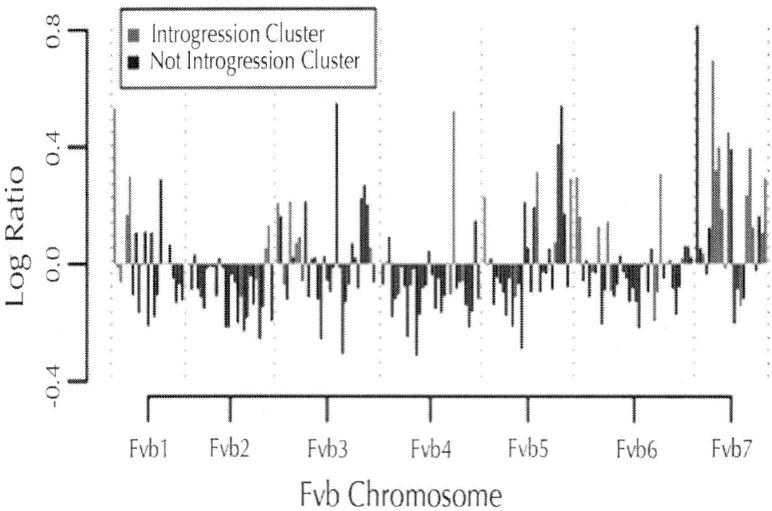

Figure 8: Ratio of *F. vesca*-like read depth to *F. iinumae*-like read depth for the two maternal octoploid parents in 1-Mb bins across the Fvb genome. Bins overlapping introgression clusters as determined from the linkage map data (red) show a much higher ratio. The ratio in all bins is likely skewed toward *F. vesca* because such reads are more likely to map to Fvb.

Interchromosome Rearrangements

We were able to assign 777 SSR markers from *F.* × *ananassa* to Fvb positions. Among the three octoploid species, we observed 139 interchromosomal rearrangements, defined as regions of Fvb in which an LG SNP maps to a linkage group pertaining to a different Fvb pseudochromosome. The 139 interchromosomal rearrangements are highly clustered, falling into only 110 1-kb regions of Fvb, 25 of which harbor at least two rearrangements, 9 of which harbor at least three rearrangements, and 1 of which harbors four rearrangements. For 120 rearrangements, we observed direct conflicts among our linkage maps, with LG SNPs from the same 1-kb region mapping to linkage groups pertaining to different Fvb pseudochromosomes. Only 25 rearrangements are supported by at least two LG SNPs, and only 4 by LG SNPs greater than 1 kb apart. Subgenome Av showed fewer rearrangements relative to the other subgenomes, consistent with its closer phylogenetic relationship to Fvb; the exceptions were regions with low mean normalized depth per targeted region in the octoploids versus Fvb-s (putatively duplicated within *F. vesca* ssp. *bracteata*), and nontargeted regions (possibly duplicated throughout *Fragaria*).

To minimize variance in coverage and standardize expected distribution patterns, we restricted further analysis to the 105 rearrangements that occur within 1 kb of one of 76 targeted regions. There are 20 targeted regions with two or more of these rearrangements, and each of the 105 of them occurs on a targeted region with 0.74 other rearrangements on average. However, if 105 rearrangements were randomly distributed among all 2,598 targeted regions with LG SNPs in the octoploids, only two regions with multiple rearrangements are expected (95% CI: 1–3), and the average rearrangement would share a targeted region with only 0.04 others (95% CI: 0.00–0.10), a highly significant difference indicating nonrandom clustering. The ratio of mean normalized depth per targeted region in the octoploids versus Fvb-s was higher and more variable at targeted regions showing rearrangements (mean ratio = 1.34, SD = 0.82) relative to other targeted regions with LG SNPs in the octoploids (mean ratio = 1.04, SD = 0.34; *t*-test, $t = 3.14$, $P < 0.01$), suggesting copy number variation caused by duplication of these regions in some lineages since the common *Fragaria* ancestor (fig. 7*B*).

DISCUSSION

Using a novel approach, POLiMAPS, we have generated a phylogeny of octoploid homeologous subgenomes. We have successfully recovered and assigned a phylogenetic position to all eight expected linkage groups for every haploid chromosome in two independent octoploid crosses (figs. 4and 5). Unlike methods based on individual genes (Senchina et al. 2003;Rousseau-Gueutin et al. 2009; Shimizu-Inatsugi et al. 2009; DiMeglio et al. 2014), we employ thousands of genome-wide markers, revealing the evolutionary history of each subgenome as a whole. POLiMAPS enhances mapping-by-sequencing (James et al. 2013) by mapping targeted sequence three ways: To a reference genome, to a linkage group, and to a phylogenetic tree, in order to highlight the evolutionary processes that produce consistencies and discrepancies among these three types of maps. Our results reveal that divergence among homeologs is heavily influenced by dynamic genomic changes affecting many sites at once, including introgression and chromosomal rearrangements.

A fundamental result of this study is that only one subgenome (Av) groups with *F. vesca* in our phylogenies, whereas three subgenomes (Bi, B1, and B2) group with *F. iinumae* (fig. 5). These data allow us to reject the prevailing hypothesis which posits two *F. vesca*-like subgenomes (Rousseau-Gueutin et al. 2009; Sargent et al. 2012). A parsimonious evolutionary history consistent with our results is that a *F. vesca*-like diploid hybridized with a *F. iinumae*-like diploid to form an allotetraploid (AvAvBiBi), which then hybridized with an unknown *F. iinumae*-like autotetraploid (B1B1B2B2) to form the octoploid ancestor of *F. chiloensis*and *F. virginiana*. Our results otherwise support the relationships among*Fragaria* species suggested by previous studies of nuclear (Rousseau-Gueutin et al. 2009) and plastome (Njuguna et al. 2013) sequence. Specifically, we confirm that *F. iinumae* is phylogenetically isolated, that *F. mandshurica* and *F. bucharica* are closely related to *F. vesca*, that the octoploids have a common origin, and that *F. vesca* ssp. *bracteata* has a close, likely conspecific relationship with one of the diploid ancestors of the octoploids (Av). For our analyses we assumed, based on previous work (Rousseau-Gueutin et al. 2009; Njuguna et al. 2013), that other *Fragaria*diploids not sampled are not candidate progenitors of the octoploids. Our results are consistent with the accepted view of disomic inheritance in which homeologous chromosomes are independent and do not cosegregate (Bringhurst 1990; Ashley et al. 2003; Rousseau-Gueutin et al. 2008). Specifically, subgenomes reliably form distinct clades, rather than multiple subgenomes from the same species forming a clade together as would be expected under free recombination among homeologs (Wendel et al. 1995). Although polysomic recombination has not eliminated the phylogenetic signal, suggesting a highly diploidized genome, a small amount of polysomic inheritance may still occur, especially between B1 and B2 (Lerceteau-Köhler et al. 2003).

With POLiMAPS, specific alleles of interest can be placed in a phylogenetic context. For example, prior work has shown that both *F. virginiana* ssp. *virginiana* and *F. chiloensis* have a Mendelian sex locus mapping to different ends of different chromosomes in homeologous group VI, suggesting either independent origins or translocation (Goldberg et al. 2010). Our results reveal that the *F. chiloensis* sex locus maps to subgenome Av whereas the *F. virginiana* ssp. *virginiana* sex locus maps to subgenome B2 The "use" of different autosomes for sex chromosomes has been seen among species in the same genera in other taxa (e.g., sticklebacks; Ross et al. 2009) but this is the first study to clearly demonstrate the use of the same chromosomes but from different ancestors. This example demonstrates the power of our phylogenetic model to enhance understanding of trait and chromosomal evolution, which will stimulate future research in strawberry genetics, including for agriculturally important traits, especially with respect to using the *F. vesca* reference genome as a proxy for cultivated strawberry.

Homoplasy consistent with introgression has been reported in many polyploid species, although it can be challenging to conclusively identify and interpret (Gaeta and Pires 2010; Wijnker et al. 2013; Chalhoub et al. 2014; Marcussen et al. 2014; Qi et al. 2014). Detection methods based on sequencing alone cannot simultaneously distinguish between gene loss and introgression, rule out homoplasy due to independent mutations, and connect distinct introgression events to a common subgenome (Buggs et al. 2009; Wang et al. 2009; Salmon et al. 2010). Although our phylogeny is well supported, a large number of sites display homoplasy (fig. 6). We identified 48 genomic clusters containing at least two regions consistent with introgression for a particular subgenome, and with few or no markers supportive of the phylogenetic position of that subgenome. There are at least two reasons why these 48 clusters are likely to represent true introgression events. First, all of them convert an *F. iinumae*-like subgenome (Bi, B1, or B2) to be more *F. vesca*-like, whereas none of them converts subgenome Av to be more *F. iinumae*-like, a highly nonrandom pattern. Second, instances of introgression homoplasy are much more common than outgroup homoplasy, even though both should occur at a similar rate if they are only due to mutation, sequencing errors, or incomplete linage sorting (Green et al. 2010). These 48 clusters represent a nontrivial portion of the haploid genome (7%), and because our conservative method has missed any introgression events that do not span two marker-containing regions, the true proportion of the genome showing introgression is likely even higher. Our whole-genome sequencing of the parents indicates that mapped *F. vesca*-like reads are at least as common as mapped *F. iinumae*-like reads in approximately 30% of the genome outside of our designated introgression regions (fig. 8). Because *F. vesca*-like reads have a higher probability of mapping to Fvb, we are reluctant to use this ratio alone to predict introgression.

Nevertheless, genomic sections outside of the introgression clusters but showing an exceptionally high proportion of *F. vesca*-like reads (fig. 8) may represent additional introgression events. Note that the failure of some *F. iinumae*-like sequence to map to Fvb is unlikely to be the cause of the unidirectional introgression pattern, as all informative markers had nonzero coverage in *F. iinumae* by definition.

Introgression could have occurred either before or after polyploidization. If it occurred before then, it would represent ancestral homoploid hybridization events among the diploid ancestors. Specifically, the *F. iinumae*-like diploids contributing to subgenomes B1, B2, and Bi would have had a small amount of *F. vesca* ancestry due to interspecies crossing followed by backcrossing to *F. iinumae*, without a change in chromosome number, prior to the subsequent allopolyploidization event. Because subgenomes B1, B2, and Bi do not form a clade to the exclusion of *F. iinumae*, this scenario requires at least two unrelated homoploid hybrid diploid ancestors. Fossil and biogeographical evidence place the octoploid ancestor in Beringia during the Pleistocene (Liston et al. 2014), and thus the extent of the original hybrid zone and the degree of homoploid introgression may not be reflected in any extant diploid population. Alternatively, introgression could have happened after polyploidization through homeologous exchange. Homeologous gene conversion, the nonreciprocal replacement of an allele with a template sequence from another subgenome, generally happens over short (<1 kb) genomic scales, but has been reported on scales greater than 1 Mb (Jacquemin et al. 2011). Reciprocal homeologous recombination, while not initially eliminating any sequence, will lead to the loss of alleles in progeny that do not inherit both recombinant chromosomes. Reciprocal homeologous recombination can cause extensive change that can rapidly destabilize the genome, which may be why this process is generally suppressed through diploidization in evolutionary successful polyploids (Gaeta and Pires 2010). Many introgression clusters are shared between the two octoploid species (fig. 4; table 2), so if they occurred after polyploidization, then they must have occurred prior to the subsequent speciation of *F. virginiana* and *F. chiloensis*.

The adaptive significance of the observed introgression and its unidirectionality is unknown. Although it is known that the degree of introgression can vary among polyploid subgenomes (Page et al. 2013;Chalhoub et al. 2014), the extreme bias observed here is remarkable and unexpected. If it occurred before polyploidization, and by chance the four diploid ancestors happened to include three homoploid hybrids (*F. iinumae* × *F. vesca*) and one *F. vesca* individual, natural selection may not have played a role. On the other hand, if such homoploid hybrids were rare, it is unlikely that all three *F. iinumae*-like subgenomes would have hybrid ancestry unless this somehow facilitated polyploid hybridization and the subsequent success of the octoploids. If introgression took place among subgenomes after polyploidization, the unidirectional pattern was likely to have been driven by some consistent evolutionary process, presumably strong selection. Although homeologous exchange can occur preferentially in one direction for mechanistic rather than adaptive reasons (as in Lange et al. 2011), this explanation seems unlikely because the genomes of *F. vesca* and *F. iinumae* are quite similar in length, GC content, and other metrics (Hirakawa et al. 2014). Regardless of when introgression occurred, we can envision two major adaptive hypotheses. First, *F. vesca* sequence per se might have higher fitness. Because *F. vesca*is the cytoplasm donor, there could have been selection for Av subgenome sequence to promote cytonuclear accommodation (Gong et al. 2012), or *F. vesca*-like genes may have been favored in the ecological conditions experienced by the octoploids. We cannot reject this hypothesis, but we note that introgression clusters mostly do not overlap among subgenomes (fig. 4), even though this would be expected if a small number of specific *F. iinumae*-like genes are deleterious in *F. vesca*-like cytoplasm or the ecological habitat of the polyploids. Second, there may have been selection to make the *F. iinumae*-like subgenomes, especially B1 and B2, more divergent from each other, in order to minimize polysomic inheritance which is often associated with lower fitness (Le Comber et al. 2010; Feldman and Levy 2012). It is noteworthy that the *F. iinumae*-like subgenomes typically display disomic inheritance and form independent clades in our phylogenies, despite a recent common origin, particularly for B1 and B2. This puzzling result may be explained by introgression causing rapid divergence among the subgenomes, and it is plausible that the adaptive value of disomic inheritance was in fact a selective advantage favoring introgression.

Genomic rearrangement in polyploids can be extensive (Lim et al. 2004;Udall et al. 2005; Chen and Ni 2006). Transpositions and translocations occurring immediately after polyploidization caused by "genomic shock" have received a great deal of attention (Lim et al. 2004; Chen and Ni 2006), but many rearrangements may also occur in diploid ancestors which are then brought together in allopolyploidy. We observed 139 interchromosome rearrangements in our linkage maps. Most of these show direct contradictions, with linkage groups pertaining to different haploid chromosomes containing LG SNPs from the same targeted region in Fvb, and therefore they cannot be explained simply as assembly errors in our reference genome. Our results add to previous observations of interchromosome rearrangements in *F. virginiana* ssp. *virginiana* detected with SSR markers (Spigler et al. 2010). Targeted regions showing rearrangements have higher and more variable depth, relative to diploids, than the rest of the genome, suggesting that homologous sequence is found in multiple locations (fig. 7*B*). In other words, many of these observations may represent a "copy-and-paste" rather than a "cut-and-paste" mechanism, consistent with the activity of Class I or Class II, Subclass 2 transposable elements (Wicker et al. 2007). Of the 110 targeted regions showing rearrangements, 22% contain LG SNPs mapping to linkage groups pertaining to at least two additional Fvb pseudochromosomes, implying repeatedly duplicated segments. After excluding regions that may be duplicated throughout *Fragaria* or just in *F. vesca* ssp. *bracteata* (nontargeted and low octoploid:diploid depth ratio), most rearrangements occur in subgenomes B1, B2, or Bi, reflecting their greater phylogenetic distance from Fvb. Thus, many of the differences between subgenomes are likely to be interchromosome rearrangements that occurred in the diploid ancestors as they diverged from each other. However, those that do occur in subgenome Av must have happened in the relatively short time since this subgenome diverged from its diploid *F. vesca*-like ancestor, suggesting that postpolyploidization rearrangements may also be important, consistent with the common observation of increased transposition in interspecific hybrids (Fontdevila 2005; Chen and Ni 2006). Because our targeted capture approach ignores the majority of the genome, and rearrangements are usually quite small and do not span between targeted regions, there are likely to be many more small rearrangements that were undetected. Our results demonstrate that translocation and/or transposition is a major contributor to differentiation between homeologous chromosomes and is consistent with the hypothesis that transposable elements are important drivers of plant evolution (Lisch 2013). Putative intrachromosome rearrangements are also apparent as changes in LG SNP order within linkage groups relative to Fvb. We did not attempt to quantify intrachromosome translocations because in many cases LG SNP order within linkage groups has low statistical support and would be affected by genotyping errors. However, one very clear inversion stands out, comprising more than half of LG II_B1 in all four octoploid parents (fig. 4). Our data are consistent with this being the same inversion previously identified in octoploid *F.* × *ananassa*(van Dijk et al. 2014).

Taken together, our results clarify previous observations of cytological and inheritance patterns in *Fragaria*. Although inheritance in octoploid*Fragaria* is primarily disomic, studies of multivalent pairings in hybrids have suggested numerous conflicting cytological models (Ichijima 1926;Yarnell 1931; East 1934; Federova 1946; Senanayake and Bringhurst 1967; Bringhurst 1990). Bringhurst (1990) synthesized the available evidence and proposed a strawberry genomic formula of AAA′A′BBB′B′, subsequently supported by single-gene phylogenies (Rousseau-Gueutin et al. 2009). Our work suggests instead a pattern of AABBB′B′B″B″, similar to the older AABBBBCC model (Federova 1946), with C (Bi) being more similar to B (B1 and B2) than A (Av). Previous *F. vesca* × *F. virginiana*crosses also support a single *F. vesca*-like subgenome, namely, Ichijima (1926) found 7 bivalent and 21 univalent chromosomes, whereas East (1934) found a hybrid in which only a single *F. virginiana* subgenome had been retained. Furthermore, Nathewet et al. (2010) found that most octoploid chromosomes had morphology resembling *F. iinumae*. However, our observation of widespread introgression means that no one pattern holds across all portions of the genome. Indeed, the nuclear gene*DHAR* used by Rousseau-Gueutin et al. (2009) for phylogenetics occurs at Fvb7_16,080,000, just 6 kb upstream from a 386-kb introgression cluster (table 2), and likely encompassed by the introgression event, as the nearest marker showing phylogenetic support is more than 100 kb farther upstream. Thus, although polyphyly of allopolyploids is known (Chalhoub et al. 2014) and cannot be ruled out as an explanation for discrepant results, the data are consistent with a monophyletic octoploid clade but with a phylogenetic signal that varies across the genome. As cytological pairings can be determined by a small number of specific genes (Sears 1976; Vega and Feldman 1998), it is perhaps not surprising to see poor concordance among cytology studies, single-gene phylogenies, and genome-wide phylogenies in many polyploids (Hancock 2012).

Indeed, the most appropriate genomic formula for a polyploid may vary depending on whether its intent is to reflect evolutionary history, overall sequence divergence, or extant segregation patterns.

Our new *Fragaria* genome assembly, Fvb, differs from the FvH4 assembly by more than 100 rearrangements (fig. 2). Some of these could represent real translocations and some could represent assembly errors in FvH4. We suspect that a majority represent assembly errors for four reasons. First, we find no inconsistencies among the three *F. vesca* ssp. *bracteata* maps, but if genome rearrangement were so rampant as to have caused all the rearrangements we see, we would expect to see at least one between *F. vesca* ssp. *bracteata* maps from different geographic locations. Second, most rearrangements were consistent with an entire scaffold being misplaced, which is a likely type of methodological error. Third, discrepancies between the octoploids and Fvb rarely support the FvH4 assembly instead, but rather are usually different from both assemblies Fourth, Fvb shows much higher synteny with the *Prunus* genome than FvH4 does (fig. 3). Nearly a third of putative translocations between FvH4 and *Prunus* are eliminated in Fvb, suggesting they are likely assembly errors in FvH4 rather than true genomic rearrangements. Our results underscore the caveat that genome assemblies are imperfect and genomic rearrangements will appear to be more common than they really are due to errors (Bhutkar et al. 2006). For example, the observation that FvH4 shows more genomic rearrangements than other eudicots including*Prunus*, *Coffea*, and *Vitis* (Illa et al. 2011; Denoeud et al. 2014) could be entirely due to assembly errors rather than biological reality. Even if some rearrangements are real differences between *F. vesca* subspecies, the closest diploid relative of the Av subgenome in the octoploids (including the cultivated *F.* × *ananassa*) is *F. vesca* ssp. *bracteata* (Njuguna et al. 2013), and therefore our Fvb assembly will be a useful resource for all strawberry geneticists.

Our approach to genome assembly and polyploid phylogenetics, based on targeted capture and dense linkage maps, will likely be fruitful when applied to other taxa. Several features of our success with this study are particularly noteworthy. First, the Fvb-s map was generated from a single selfed individual, contradicting the common assumption that useful linkage maps can only be generated by crossing distantly related parents. Second, our targeted capture probes were designed to target polymorphisms in Fvb-m and Fvb-p (Tennessen et al. 2013), yet they yielded many useful polymorphisms in Fvb-s and the octoploids, indicating that for sufficiently heterozygous individuals, there is no need to identify polymorphisms in advance. Third, our success with calling segregating SNPs in polyploids suggests that many of the challenges in polyploid genetics (Dufresne et al. 2014) are ameliorated when sequencing coverage is high enough such that variants can be accurately identified even with relatively simple algorithms. Fourth, our novel POLiMAPS method of identifying phylogenetic markers that share short sequencing reads with known linkage map markers will allow otherwise intractable polyploid genomes to be dissected (Hirsch and Buell 2013). Testing for the major patterns described here in other taxa will illuminate the complex evolutionary processes associated with whole-genome duplications.

ACKNOWLEDGMENTS

The authors thank Katie Schuller and Matt Koski for assistance with crosses and plant cultivation, Chris Edwards and Kevin Weitemier for libraries and target enrichment, Brian Husband and the USDA ARS National Clonal Germplasm Repository (Corvallis, Oregon) for providing plant samples, Mark Dasenko and the OSU CGRB for sequencing, Sanjuro Jogdeo for data curation, and Rich Cronn for discussion. This work was supported by the National Science Foundation to T.L.A. (DEB 1020523 and DEB 1241006) and A.L. (DEB 1020271 and DEB 1241217), and by the United States Department of Agriculture to T.L.A. and the University of Pittsburgh (USDA-CSREES 2005-00765).

REFERENCES

1. Ashley MV, et al High variability and disomic segregation of microsatellites in the octoploid *Fragaria virginiana* Mill. (Rosaceae). Theor Appl Genet. 2003; 107:1201-1207.
2. Bhutkar A, Russo S, Smith TF, Gelbart WM Techniques for multi-genome synteny analysis to overcome assembly limitations. Genome Inform. 2006; 17:152-161.

3. Bringhurst RS Cytogenetics and evolution in American *Fragaria*. HortScience1990; 25:879-881.
4. Buggs RJ, et al Gene loss and silencing in *Tragopogon miscellus* (Asteraceae): comparison of natural and synthetic allotetraploids. Heredity 2009; 103:73-81.
5. Byrne D, Jelenkovic G Cytological diploidization in the cultivated octoploid strawberry *F.* × *ananassa*. Can J Genet Cytol. 1976; 18:653-659.
6. Cenci A, Combes MC, Lashermes P Genome evolution in diploid and tetraploid*Coffea* species as revealed by comparative analysis of orthologous genome segments.Plant Mol Biol. 2012;78:135-145.
7. Chalhoub B, et al Early allopolyploid evolution in the post-Neolithic *Brassica napus*oilseed genome. Science 2014; 345:950-953.
8. Chen ZJ, Ni Z Mechanisms of genomic rearrangements and gene expression changes in plant polyploids. Bioessays 2006; 28:240-252.
9. Denoeud F, et al The coffee genome provides insight into the convergent evolution of caffeine biosynthesis. Science 2014; 345:1181-1184.
10. DiMeglio LM, Staudt G, Yu H, Davis TM
11. . A phylogenetic analysis of the genus*Fragaria* (strawberry) using intron-containing sequence from the *ADH*-1 gene. PLoS One2014; 9:e102237.
12. Dufresne F, Stift M, Vergilino R, Mable BK Recent progress and challenges in population genetics of polyploid organisms: an overview of current state-of-the-art molecular and statistical tools. Mol Ecol. 2014; 23:40-69.
13. Doyle JJ, Doyle JL A rapid DNA isolation procedure for small quantities of fresh leaf tissue. Phytochem Bull. 1987; 19:11-15.
14. East EM A novel type of hybridity in *Fragaria*. Genetics 1934; 19:167-174.
15. Federova NJ Crossability and phylogenetic relations in the main European species of *Fragaria*. Compte Rend (Doklady) Acad Sci USSR. 1946; 52:545-547.

16. Feldman M, Levy AA Genome evolution due to allopolyploidization in wheat.Genetics 2012; 192:763-774.
17. Ferrarini M, et al An evaluation of the PacBio RS platform for sequencing and de novo assembly of a chloroplast genome. BMC Genomics 2013; 14:670.
18. Fontdevila A Hybrid genome evolution by transposition. Cytogenet Genome Res.2005; 110:49-55.
19. Gaeta RT, Pires JC Homoeologous recombination in allopolyploids: the polyploid ratchet. New Phytol. 2010; 186:18-28.
20. Goldberg MT, Spigler RB, Ashman TL Comparative genetic mapping points to different sex chromosomes in sibling species of wild strawberry (*Fragaria*). Genetics2010; 186:1425-1433.
21. Gong L, et al The cytonuclear dimension of allopolyploid evolution: an example from cotton using rubisco. Mol Biol Evol. 2012; 29:3023-3036.
22. Green RE, et al A draft sequence of the Neandertal genome. Science2010; 328:710-722.
23. Hancock JF Plant evolution and the origin of crop species. 2012. Wallingford (United Kingdom): CABI.
24. Hirakawa H, et al Dissection of the octoploid strawberry genome by deep sequencing of the genomes of *Fragaria* species. DNA Res. 2014; 21:169-181.
25. Hirsch CN, Buell CR Tapping the promise of genomics in species with complex, nonmodel genomes. Annu Rev Plant Biol. 2013; 64:89-110.
26. Hollister JD, et al Genetic adaptation associated with genome-doubling in autotetraploid *Arabidopsis arenosa*. PLoS Genet. 2012; 8:e1003093.
27. Hyun TK, et al De-novo RNA sequencing and metabolite profiling to identify genes involved in anthocyanin biosynthesis in Korean black raspberry (*Rubus coreanus*Miquel). PLoS One 2014; 9:e88292.
28. Ichijima K Cytological and genetic studies on *Fragaria*. Genetics 1926; 11:590-604.
29. Illa E, et al Comparative analysis of rosaceous genomes and the reconstruction of a putative ancestral genome for the family. BMC Evol Biol. 2011; 11:9.

30. International Peach Genome Initiative, et al. The high-quality draft genome of peach (*Prunus persica*) identifies unique patterns of genetic diversity, domestication and genome evolution. Nat Genet. 2013; 45:487-494.
31. Isobe SN, et al Construction of an integrated high density simple sequence repeat linkage map in cultivated strawberry (*Fragaria × ananassa*) and its applicability. DNA Res. 2013; 20:79-92.
32. Jacquemin J, et al Long-range and targeted ectopic recombination between the two homeologous chromosomes 11 and 12 in Oryza species. Mol Biol Evol. 2011; 28:3139-3150.
33. James GV, et al User guide for mapping-by-sequencing in *Arabidopsis*. Genome Biol. 2013; 14(6):R61.
34. Jiao Y, et al Ancestral polyploidy in seed plants and angiosperms. Nature2011; 473:97-100.
35. Kovarik A, et al Concerted evolution of 18-5.8-26S rDNA repeats in *Nicotiana allotetraploids*. Biol J Linn Soc Lond. 2004; 82:615-625.
36. Kent WJ BLAT—the BLAST-like alignment tool. Genome Res. 2002; 12:656-664.
37. Lange C, Zerulla K, Breuert S, Soppa J Gene conversion results in the equalization of genome copies in the polyploid haloarchaeon *Haloferax volcanii*. Mol Microbiol.2011; 80:666-677.
38. Le Comber SC, Ainouche ML, Kovarik A, Leitch AR Making a functional diploid: from polysomic to disomic inheritance. New Phytol. 2010; 186:113-122.
39. Lerceteau-Köhler E, Guérin G, Laigret F, Denoyes-Rothan B Characterization of mixed disomic and polysomic inheritance in the octoploid strawberry (*Fragaria × ananassa*) using AFLP mapping. Theor Appl Genet. 2003; 107:619-628.
40. Li H, Durbin R Fast and accurate short read alignment with Burrows-Wheeler transform. Bioinformatics 2009; 25: 1754-1760.
41. Li H, et al The sequence alignment/map (SAM) format and SAMtools. Bioinformatics2009; 25: 2078-2079.
42. Lim KY, Matyasek R, Kovarik A, Leitch AR Genome evolution in allotetraploid*Nicotiana*. Biol J Linn Soc Lond. 2004; 82:599-606.

43. Lipman MJ, Chester M, Soltis PS, Soltis DE Natural hybrids between *Tragopogon mirus* and *T. miscellus* (Asteraceae): a new perspective on karyotypic changes following hybridization at the polyploid level. Am J Bot. 2013; 100:2016- 2022?

44. Lisch D How important are transposons for plant evolution? Nat Rev Genet.2013; 14:49-61.

45. Liston L, Cronn R, Ashman T-L *Fragaria*: a genus with deep historical roots and ripe for evolutionary and ecological insights. Am J Bot. 2014; 101:1686-1699?

46. Madlung A, et al Genomic changes in synthetic *Arabidopsis polyploids*. Plant J.2005; 41: 221-230.

47. Maluszynska J, Hasterok R Identification of individual chromosomes and parental genomes in *Brassica juncea* using GISH and FISH. Cytogenet Genome Res.2005; 109:310-314.

48. Marcussen T, et al Ancient hybridizations among the ancestral genomes of bread wheat. Science 2014; 345:1250092.

49. Margarido GRA, Souza AP, Garcia AAF OneMap: software for genetic mapping in outcrossing species. Hereditas 2007; 144:78-79.

50. Nathewet P, Hummer KE, Yanagi T, Iwatsubo Y, Sone K Karyotype analysis in octoploid and decaploid wild strawberries in *Fragaria* (Rosaceae). Cytologia2010; 75:277-288.

51. Njuguna W, Liston A, Cronn R, Ashman T-L, Bassil N Insights into phylogeny, sex function and age of *Fragaria* based on whole chloroplast genome sequencing. Mol Phylogenet Evol. 2013; 66:17-29.

52. Otto SP, Whitton J Polyploid incidence and evolution. Annu Rev Genet.2000; 34:401-437.

53. Page JT, et al Insights into the evolution of cotton diploids and polyploids from whole-genome re-sequencing. G3 2013; 3: 1809-1818.

54. Qi J, Chen Y, Copenhaver GP, Ma H Detection of genomic variations and DNA polymorphisms and impact on analysis of meiotic recombination and genetic mapping.Proc Natl Acad Sci U S A. 2014;111:10007-10012.

55. Ross JA, Urton JR, Boland J, Shapiro MD, Peichel CL Turnover of sex chromosomes in the stickleback fishes (gasterosteidae). PLoS Genet. 2009; 5:e1000391.
56. Rousseau-Gueutin M, et al Comparative genetic mapping between octoploid and diploid *Fragaria* species reveals a high level of colinearity between their genomes and the essentially disomic behavior of the cultivated octoploid strawberry. Genetics2008; 179:2045-2060.
57. Rousseau-Gueutin M, et al tracking the evolutionary history of polyploidy in*Fragaria* L. (strawberry): new insights from phylogenetic analyses of low-copy nuclear genes. Mol Phylogenet Evol. 2009; 51:515-530.
58. Salmon A, Flagel L, Ying B, Udall JA, Wendel JF Homoeologous nonreciprocal recombination in polyploid cotton. New Phytol. 2010; 186:123-134.
59. Sargent DJ, et al Simple sequence repeat marker development and mapping targeted to previously unmapped regions of the strawberry genome sequence. Plant Genome 2011; 4:165-177.
60. Sargent DJ, et al A microsatellite linkage map for the cultivated strawberry (*Fragaria × ananassa*) suggests extensive regions of homozygosity in the genome that may have resulted from breeding and selection. Theor Appl Genet. 2012; 124:1229-1240.
61. Sears ER Genetic control of chromosome pairing in wheat. Annu Rev Genet.1976; 10:31-51.
62. Senanayake YDA, Bringhurst RS Origin of *Fragaria* polyploids. I. Cytological analysis.Am J Bot. 1967; 54:221-228.
63. Senchina DS, et al Rate variation among nuclear genes and the age of polyploidy in*Gossypium*. Mol Biol Evol. 2003; 20:633-643.
64. Shimizu-Inatsugi R, et al the allopolyploid *Arabidopsis kamchatica* originated from multiple individuals of *Arabidopsis lyrata* and *Arabidopsis halleri*. Mol Ecol.2009; 18:4024-4048.
65. Shulaev V, et al The genome of woodland strawberry (*Fragaria vesca*). Nat Genet.2011; 43:109-116.

66. Spigler RB, Lewers KS, Ashman T-L Genetic architecture of sexual dimorphism in a subdioecious plant with a proto-sex chromosome. Evolution 2011; 65:1114-1126.

67. Spigler RB, Lewers KS, Johnson AL, Ashman T-L Comparative mapping reveals autosomal origin of sex chromosome in octoploid *Fragaria virginiana*. J Hered.2010; 101:S107.

68. Spigler RB, Lewers KS, Main DS, Ashman T-L Genetic mapping of sex determination in a wild strawberry, *Fragaria virginiana*, reveals earliest form of sex chromosome.Heredity 2008;101:507-517.

69. Stamatakis A RAxML-VI-HPC: maximum likelihood-based phylogenetic analyses with thousands of taxa and mixed models. Bioinformatics 2006; 22:2688-2690.

70. Tennessen JA, Govindarajulu R, Liston A, Ashman TL Targeted sequence capture provides insight into genome structure and genetics of male sterility in a gynodioecious diploid strawberry, *Fragaria vesca* ssp. *bracteata* (Rosaceae). G3 2013; 3:1341-1351.

71. Udall JA, Quijada PA, Osborn TC Detection of chromosomal rearrangements derived from homologous recombination in four mapping populations of *Brassica napus*L. Genetics 2005; 169:967-979.

72. Van Dijk T, et al Genomic rearrangements and signatures of breeding in the allo-octoploid strawberry as revealed through an allele dose based SSR linkage map. BMC Plant Biol. 2014; 14:55.

73. Vega JM, Feldman M Effect of the pairing gene Ph1 on centromere misdivision in common wheat. Genetics 1998; 148:1285-1294.

74. Wang X, Tang H, Bowers JE, Paterson AH Comparative inference of illegitimate recombination between rice and sorghum duplicated genes produced by polyploidization. Genome Res. 2009; 19:1026-1032.

75. Wendel JF, chnabel A, Seelanan T Bidirectional interlocus concerted evolution following allopolyploid speciation in cotton (*Gossypium*). Proc Natl Acad Sci U S A.1995; 92:280-284.

76. Wicker T, et al A unified classification system for eukaryotic transposable elements.Nat Rev Genet. 2007; 8:973-982.

77. Wijnker E, et al The genomic landscape of meiotic crossovers and gene conversions in *Arabidopsis thaliana*. Elife 2013; 2: e01426.
78. Wood TE, et al The frequency of polyploid speciation in vascular plants. Proc Natl Acad Sci U S A. 2009; 106:13875-13879.
79. Woodhouse MR Following tetraploidy in maize, a short deletion mechanism removed genes preferentially from one of the two homeologs. PLoS Biol.2010; 8: e1000409.
80. Yarnell SH Genetic and cytological studies on *Fragaria*. Genetics 1931; 16:422-454.
81. Xiong Z, Pires JC Karyotype and identification of all homoeologous chromosomes of allopolyploid *Brassica napus* and its diploid progenitors. Genetics 2011; 187:37-49.

Chapter 3

Plant Domestication and Resistance to Herbivory

Bhupendra Chaudhary

School of Biotechnology, Gautam Buddha University, Greater Noida 201 308, India

ABSTRACT

Transformation of wild species into elite cultivars through "domestication" entails evolutionary responses in which plant populations adapt to selection. Domestication is a process characterized by the occurrence of key mutations in morphological, phenological, or utility genes, which leads to the increased adaptation and use of the plant; however, this process followed by modern plant breeding practices has presumably narrowed the genetic diversity in crop plants. The reduction of genetic diversity could result in "broad susceptibility" to newly emerging herbivores and pathogens, thereby threatening long-term crop retention. Different QTLs influencing herbivore resistance have also been identified, which overlap with other genes of small effect regulating resistance indicating the presence of pleiotropism or linkage between such genes. However, this reduction in genetic variability could be remunerated by introgression of novel traits from wild perhaps with antifeedant and antinutritional toxic components. Thus it is strongly believed that transgenic technologies may provide a radical and promising solution to combat herbivory as these avoid linkage drag and also the antifeedant angle. Here, important questions related to the temporal dynamics of resistance to herbivory and intricate genetic phenomenon with their impact on crop evolution are addressed and at times hypothesized for future validation.

INTRODUCTION

During speciation in crop plants, many morphological changes evolved in response to continuous selection pressure. Such characters are largely governed by genetic and epigenetic changes or are modulated according to ecological adaptations. The transition of wild progenitor species into modern elite cultivars through "domestication" entails evolutionary responses in which plant populations adapt to human selection. In response to this selection most plant species exhibit marked changes in a variety of phenotypes, most noticeably in traits consciously under selection (e.g., fruit size, yield, and evenness of maturation) [1]. As Darwin [2] profoundly recognized long ago, the study of the phenotypic variation between wild and domesticated plants presents an opportunity to generate insight into general principles of evolution, using the morphologically variable antecedent and descendant taxa.

An example of how this concept has transformed our understanding is the realization that natural selection pressure, as well as adaptation under human selection, often led to unexpected and unexplained departures from predicted phenotypes. This mainly includes traits such as enhanced yield, enhanced apical dominance, reduced seed dormancy, perennial to annual habit, and relative susceptibility to pathogens, disease, and insect pests [3, 4]. However, the latter received the least attention during the process of "agricultural evolution." The term "agricultural evolution" here, in fact, summarizes all of the changes accumulated in any wild plant form under natural selection, human-mediated artificial selection (=domestication), and modern breeding practices (Figure 1). From an evolutionary standpoint, these phenomena may be viewed as novel generators of variation in the tertiary gene pool comprised of domesticated and wild germplasms (Figure 1). Such variations occurred mostly at genetic level and provide the ability for a given species to evolve in response to the changing environmental conditions and stress factors [5, 6]. Notwithstanding the striking discoveries of the genetic basis of evolved morphological traits in crop plants [7–11], relatively little is understood about the manner in which gene networks and biological processes are associated with the more susceptible phenotypes of the elite forms.

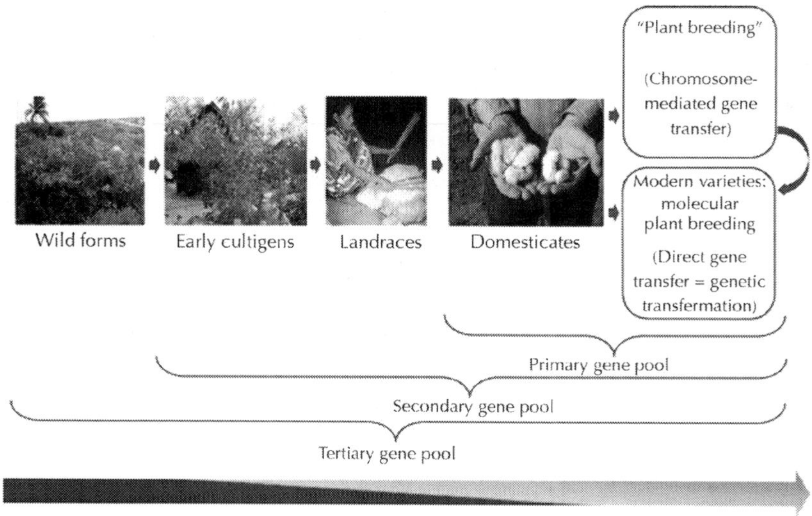

Figure 1: An example of cotton (Gossypium) evolution under human selection and contemporary breeding programs. The modern "crop" plants are the outcome of recurrent selection on wild form undergoing through early cultigens and landraces. In conventional and molecular breeding programs, it is possible to distinguish between primary, secondary, and tertiary gene pools and exchange of hereditary material. Each primary gene pool comprises one domesticated species together with those species with which it readily cross-breeds. The secondary gene pool includes species that can be cross-bred only with difficulty. The tertiary gene pool comprises those species which can be cross-bred only by using advanced techniques such as embryo rescue. (Courtesy Jonathan F. Wendel, ISU). The horizontal bar shows the reduction in genetic diversity along with domestication steps with the help of dark to lighter shades.

Regardless of these rapidly accumulating insights, important questions remain about every stage of agricultural development. How did individual crop plants evolve from wild species and acquire agriculturally important traits? How do contemporary plant forms achieve diverse evolutionary trajectories separate from those of their progenitor(s)? How do recently formed elite accessions of crop plants become compromised of different resistance traits? To what degree has crop evolution via the process of domestication and concurrent breeding practices provided a stimulus for sustainable agriculture? Despite the domestication events followed by breeding practices across plant taxa, we do not know why agricultural evolution is so prevalent or conversely, why crop evolution is not universal if it confers some adaptive advantages and promotes species diversification. Nor do we understand the dynamics that underlie the transformation of wild forms to domesticated forms in cryptic crops such as cotton and corn.

It has been assumed that several agriculturally important traits such as resistance to abiotic and biotic stress conditions decreased significantly during evolution. For example, in domesticated accessions of the genusCajanus, reduced levels of resistance have been reported against herbivores [12, 13], bacterial blight [14], and fungal diseases [15]. Among stress conditions, a reduction in drought tolerance, resistance to herbivory and pathogens, is the major threat to crop plants. It is difficult to understand what are the precise genetic underpinnings are that make a plant species vulnerable to drought, herbivores, and pathogens after passing through the evolutionary important mechanism of crop domestication? Surprisingly, what is the extent of reduction in resistance traits across crop plants, if the reduction in any particular resistance trait is proportional to another resistance trait? The answer to these questions may not be consistent across plant taxa but may only hold true for a particular plant lineage. Generally, domestication promotes heterozygosity leading to the more successful variants under selection pressure either through fixed hybridity or by polysomic inheritance. Could it be assumed that domesticated accessions are in general more "successful" than their wild progenitors? This is an exceedingly difficult question to answer, in large part because "success" is an ill-defined term that can refer to anything from short-term proliferation of individuals to long-term effects on lineage diversification. The susceptible nature of modern crop plant varieties in comparison to their wild progenitors could be one of the most apparent consequences of such a megaevent. Or accelerated mutational activity in coresident genomes (in case of polyploid crops) in early generations led to a downgrade in the pathogenic and herbivore resistance of domesticated plant species. Answering these and other questions will require comparisons of wild and domesticated forms by researchers from diverse disciplines such as ecology, population biology, and physiology. Unfortunately, these important areas of biology have received far less attention than the genetics and genomics of selection [7, 16–18]. Nevertheless, even in these better-studied areas much remains to be learned, and it is only by moving beyond the idiosyncrasies of a handful of model crop systems that emergent properties of selection can be detected. Though crop plants are threatened by many stress conditions, most of the previous research has been focused on the study of genetic end of resistance to herbivory at the gene expression level, during the process of selection, and at the phenotypic level. The aim of this paper is to provide a broad, updated entry to the literature as well as to highlight the major unanswered questions in the field of crop evolutionary genetics.

EVOLUTION OF CROP PLANTS

A primary concern of agricultural evolution biology is to investigate where, when, and how crop plants originated. Vavilov's center of domestication [19] has been a valuable hypothesis as to where crops originated and where our sessile, agrarian cultures began. Since then, we have made great strides in pinpointing where contemporary domesticated forms have arisen and from which wild species they are derived (Figure 2). It has been believed that modern elite plant varieties with useful characters (mostly hybrids) for sustainable agriculture have been developed through (i) domestication and (ii) following research and breeding activities that were implemented by scientists and breeders worldwide (Table 1) (reviewed by [20, 21]). However, it is essential to understand which has been the foremost drive largely in the phenomenon of "agricultural evolution."

Table 1: Some centers of origin of crop domestication and the trait under selection

Crop	Area of origin	Traits influenced under domestication	Source
Cereals			
Rice	China	Reduction in grain shattering and seed dormancy; synchronization of seed maturation; reduction in tiller number; increase in tiller erectness; increase in panicle branches; Number of spikelets per panicle; reduction in hull and pericarp coloration and awn length	[10, 22,23]
Barley	Fertile crescent, andIsrael-Jordan area	Reduction in grain shattering; separation of seeds from hulls	[9]
Wheat	Southwest Asia (fertile crescent)	Reduction in shattering of grains (nonbrittle rachis); free-threshing trait	[27–29]
Maize	Mesoamerica	Increased apical dominance; production of seeds in relatively large numbers	[7, 8]
Brassicas			
Cabbage	—	Large number of leaves surrounding the terminal bud	[30]
Cauliflower	—	Formation of inflorescence meristems	[31]
Legumes			

Lentil	Mesoamerica	Seed dormancy	[32]
Vigna	Southeast Asia	Increase in seed and pod size, nontwining growth habit, loss of seed dormancy, and seed dispersal ability	[24]
Pea	Southwest Asia (fertile crescent)	Indehiscent pods; lack of dormancy dwarfness; less basal branches; large seeds; good seed qualityday neutral flowering	[33]
Fibers			
Cotton	Mexico and Peru	Fiber length and quality	[16, 25]
Vegetables			
Tomato	Mesoamerica	Fruits' size, shape, and structure	[4, 26]
Potato	Andes and Amazonia	Shorter stolons, larger tubers, (often) colored and variously shaped tubers, and reduction of bitter tuber glycoalkaloids	[34]
Squash	Mesoamerica	increased seed length and peduncle diameter, change in fruit shape and color	[35]

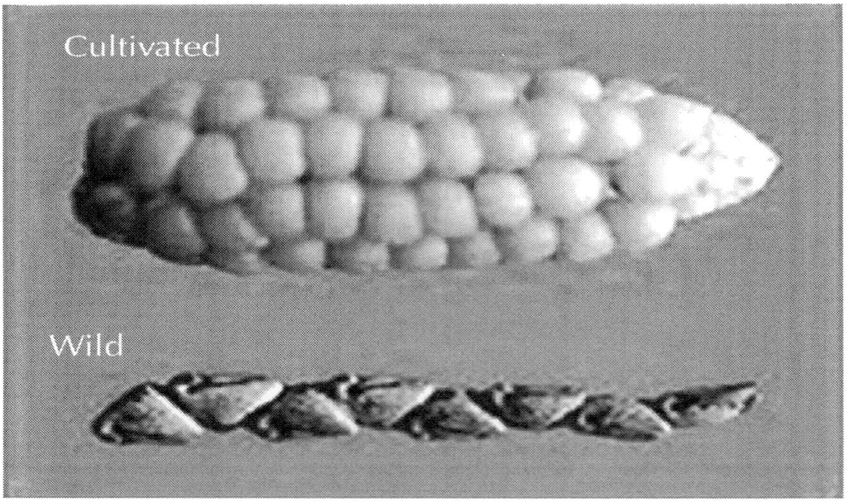

(a)

(b)

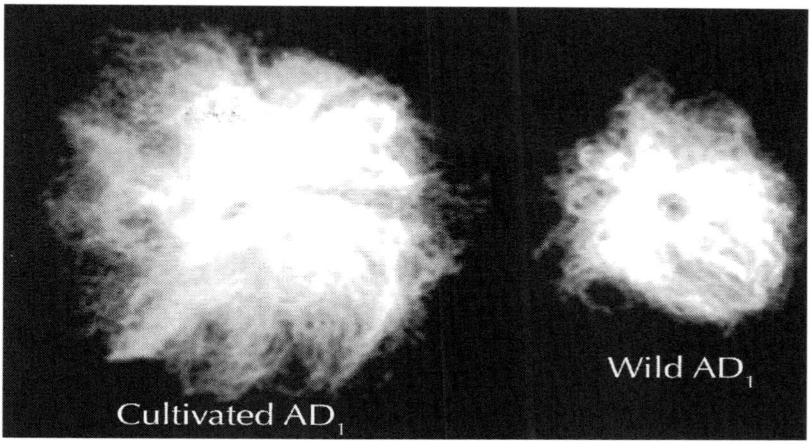

(c)

Figure 2: Difference in wild forms and their respective domesticated forms showing the impact of natural selection during evolution and domestication on phenotypic traits. (a) Evolution of maize; the domesticated maize (up) and wild teosinte (down). Teosinte has many lateral branches, while today's maize is unbranched. (b) Evolution of tomato; the much larger fruit (right) is from the domesticated Solanum lycopersicum, and the small tomato fruit (right) is from wild species Solanum pimpinellifolium; and (c) evolution of cotton (Gossypium hirsutum); the long fiber phenotype is from domesticated cotton G. hirsutum (AD_1) (left) and small fuzzy phenotype is from the wild species (right).

The term "domestication" is often used to describe the process by which wild becomes stabilized. From the standpoint of morphological transformation, domesticated forms are by definition wild species with certain traits highlighted under human selection (Figure 1), showing character modifications including novel trait formation and subsequent segregation, for example, a reduction in grain shattering and seed dormancy in rice [10, 22, 23]; an increase in seed and pod size in Vigna [24]; an increase in fiber length and quality in cotton [16,25]; changes in fruit size and shape in tomato [26]; increased apical dominance in maize [7, 8]; and more. As Darwin recognized [2], the study of the phenotypic variation between wild and domesticated plants presents an opportunity to generate insight into general principles of evolution, using morphologically variable antecedent and descendant in a comparative framework. This approach provides an intriguing perspective on the molecular genetics of human-mediated artificial selection. It is thus assumed that strong artificial selection coupled with introgression (=crossing with the respective wild relative) could drive the fixation of the most beneficial genes and their expression regulation in the process of crop domestication.

The domestication process and introgression under modern breeding programs must have served as effective means to increase the genetic diversity of elite cultivars especially following the initial domestication bottleneck, and to produce cultivars adaptive to climatic conditions [36]. But is it true for all domesticated crop plants? Can it be assumed that cereals with divergent genomic backgrounds experienced one or more domestication events in their evolutionary history? To test this hypothesis, metagenome studies of population genetic structures of cereals, other important modern taxa and their wild progenitors of known origin, are required. Instances of multiple independent domestications in cereals do not provide evidence for their clear ancestry (e.g., Oryza, Hordeum), and even if their respective wild progenitors were identified, the multiple origins of domesticated forms and their formation would mostly remain unknown. It has also been assumed that almost all crop plants have experienced repeated polyploidization of genomes in their evolutionary history with multiple-fold duplication of ancestral angiosperm (flowering plant) genes [37]. Polyploidy has influenced flowering plant diversification and provides raw material for the evolution of novelty by relaxing purifying selection on duplicated genes. However, following polyploidization, is the formation of multiple and independent domesticated forms a synchronized event across taxa? It still remains unclear whether genetic polyploids such as wheat, unlike rice and barley, must necessarily undergo mutation in orthologous loci simultaneously [38, 39]. In fact, many domesticated plant forms appear to undergo one or more mutations in a single gene, and thus the probability of parallel and independent selection of orthologous chromosomal regions responsible for the key domestication transition is expected to be weak.

Domestication could, therefore, include all responses to plant evolution, including genetic and epigenetic effects, as reviewed recently [1, 26]. What is the postdomestication impact on the genomic architecture of a plant? The domesticated forms are expected to undergo many of the genome-level megachanges both at structural and functional levels, such as sequence loss, structural rearrangements, and changes in the regulatory sequences, respectively [40, 41]. Recent large-scale microarray studies on the comparison of wild and domesticated forms of selective plant species confirmed that global gene expression had been radically altered by domestication [16, 25]. Such changes have certainly played an important role at the evolutionary scale as domesticated plant forms are not always achieved immediately, even in relatively simpler genomes, as reviewed by [17, 36]; and inheritable changes can only occur often through mutations at genic and/or regulatory levels. For example, a QTL (sh4) is responsible for the reduction of grain shattering in the wild rice [10] and a loss of function mutation of Vrs1 in six-rowed barley [9]; and a QTL is responsible for free-thrashing character in wheat [27]. Accordingly, what is the preference for such target genes to undergo mutations during selection process? It is now evident that the genes involved in important domestication transitions are preferably regulatory sequences whose mutations can generate substantial phenotypic modifications serving as suitable targets for strong artificial selection in the key steps of crop evolution [42,43].

Regarding the magnitude of the changes that occurred throughout crop formation, it appears that breeding activities of crop plants are relatively of less radical change than the conversion of wild forms to the domesticated forms. If so, will the modern elite varieties derived through hybridization of two similar domesticated varieties/forms have less evolutionary resolution than those derived through hybridization with their wild progenitor species? This is definitely an affirmative assertion and has also been validated across plant taxa with reduced genetic diversity in the breeding programs. Thus, what may be the potential risks involved with such radical loss in the genetic diversity among elite cultivars? It may also develop the weedy competitors of crop plants as well as their susceptibility to the diseases, pathogens, and herbivores leading to severe crop losses. But is there any correlation between operative stress condition and the niche of any plant population? Are there sufficient pieces of evidence for relatively less epidemic of any particular pathogen or pest population? The bewildering possibility of such prevalence may destabilize the crop productivity as well as subsequent evolution. For example, despite low crop diversity among cereals initially, successful introgression of resistance to abiotic stress conditions, pathogens, and herbivores was deployed to maintain yield. In this scenario, what is the deadline for such incorporation of resistance traits in crops for sustainable agriculture? The answer is still unclear but it may only be possible if (i) domestication, (ii) change from traditional landraces to modern breeding varieties, and (iii) their over and above decade field adaptation can work indefinitely; for example, maize hybrids in the United States now have a useful lifetime of about 4 years, half of what it was 30 years ago [44].

The advent of genomics has brought a bonafide improvement to the study of such regulatory regions and generation of molecular and expression data, knowledge, and tools which could be applied in modern breeding programs for exploitation of genes from tertiary gene pool (Figure 1). To further understand the genetic basis of domestication, tremendous variations have been revealed using molecular markers. For example, in tomato, the genetic variation present in wild species has been investigated intensively for specific traits and is being exploited for tomato breeding [45–48]. Using DNA technologies, the diversity of domesticated tomato is estimated to comprise <5% of the genetic variation as compared to the rich reservoir in wild relatives [49]. So far, even using sensitive molecular markers, very low polymorphism within the domesticated tomato gene pool has been identified, [50, 51]. A loss of genetic diversity revealed through sequence repeat markers was also observed from wild Triticum tauschii to the landraces and subsequently to the elite germplasm [40]. Though the successful application of breeding programs has produced high-yielding crop varieties, ironically the plant breeding processes have threatened the genetic basis upon which the breeding depends. If assumed so, is the redundancy in the loss of diversity across taxa deleterious, or neutral? What is the spectrum of consequences of having genes and mutations underlying domestication transitions (colloquially called "domestication syndrome")? Are the answers to these questions consistent among plant lineages or between parallel domestication events? Nevertheless, intensive artificial selection does not inevitably lead to a loss of genetic diversity, and diversity can be compensated by the introgression of novel germplasm. One example of this is the development of novel rice variety [10], with the introgression of major QTL, sh4, responsible for the reduction of grain shattering.

RESISTANCE TO HERBIVORY

Resistance Variations Arising by Domestication

Herbivores (insect pests) are the major factor responsible for destabilizing crop productivity in agricultural ecosystems. Herbivores have been recognized as a major constraint to crop production causing significant yield loss and quality degradation. In response to herbivory, plants have acquired inherent resistance against such pests; however, the intensity of this resistance varies enormously between wild and domesticated forms. So, what could be the rationale for such large variations in the resistance levels? It is believed that the alterations in herbivory resistance are a prominent outcome of genetic reduction within and between crop species. Several studies have shown that massive expression changes accompany crop evolution (under domestication and breeding practices). The magnitude of expression changes varied greatly between species influencing genetic diversity and resistance levels tremendously. However, do enough data exist at present to reveal general trends of reduced resistance to herbivory among domesticated forms across taxa? It is evident in three wild relatives of chickpeas that are, Cicer microphyllum, C. canariense, and C. macracanthus exhibiting significant reduction in leaf feeding, larval survival, and larval weight of neonate larvae of Helicoverpa armigera in comparison to domesticated perennial accessions. The extent of suppression of damage is impressive in wild accessions as compared to the domesticated chickpeas.

Also, a comparison of global gene expression profiles in the wild and domesticated allotetraploid cottonGossypium barbadense has specifically addressed the transcriptional effects of domestication during development. Several important genes and their functional categories related to resistance genes have been identified as downregulated in the domesticated form than its counterpart wild form supporting the assumption of reduced resistance properties of a plant species [16]. Furthermore, a comparison of wild and domesticated accessions of allopolyploid G. barbadense with their diploid progenitors (G. arboreum and G. raimondii; B. Chaudhary & J. F. Wendel, unpublished) revealed significant transcriptional downregulation of resistance-related genes in the expression phenotype of the domesticated form. This is a clear indication of reduction in herbivore resistance traits during the domestication of an allopolyploid crop. Since cotton allopolyploid species carry "A" and "D" genomes, derived from their diploid progenitors [52], it may also be argued that such a reduction in herbivory resistance may be the cumulative reduction occurred during both polyploid formation and domestication. However, following polyploid formation, many duplicated genes undergo transcriptional biases and do not behave as simple additive combinations of the parental genomes [53], but instead are maintained at the ancestral resistance levels at least in the wild forms. Thus, the possibility of reduction in resistance levels during polyploid formation appears to be minimal and major reductions in resistance levels may have occurred during the domestication process.

This lack of additivity in gene expression levels in most polyploid crop plants raises several fundamental questions on the consequences of the evolutionary dynamics of resistance gene expression following domestication. From a mechanistic standpoint, what is responsible for nonadditivity in gene expression, and why does this vary so much among resistance genes and between different genomic combinations? Why do such important genes from two diploid genomes demonstrate such a large disparity in the degree of suppression of the gene expression phenotype? Also from an evolutionary point of view, how does genomic evolution impact resistance gene expression variation during agricultural evolution (domestication followed by the breeding practices), and what are the potential phenotypic effects of each of these sources of variation?

Genetics of Resistance to Herbivory

One of the most recent and spectacular revelations in crop plants is the identification of a number of molecular markers, and how these markers could be applied to identify and track target genes in a marker-assisted breeding program [54–56]. Molecular markers have also been applied to increase understanding of the mechanistic and biochemical basis of herbivore resistance, as shown thoroughly in maize [57], mungbean [58], potato [59], and in soybean [60]. What are the putative regulatory genomic components and pathways providing resistance against herbivores, and at incredibly varied levels? Five QTLs have been identified inArabidopsis known to regulate the glucosinolate-myrosinase system controlling the generalist herbivoreTrichoplusia ni than specialist feeding insect Plutella xylostella [61] identified five QTLs in Arabidopsis known to regulate the glucosinolate-myrosinase system controlling the generalist herbivore Trichoplusia ni than specialist feeding insect Plutella xylostella. This demonstration of the higher levels of genetic variation for resistance to the generalist and specialist herbivores has been further verified and expanded in several subsequent studies, including one in which several QTLs from consistent resistance sources for leaf feeding insects SWCB and FAW were mapped on and observed to be located on chromosomes 6, 7, and 9 in corn. Given the resistance to both of these insects, candidate genes were identified as mir cysteine proteinase gene family [62] and the Glossly15 gene controlling adult to juvenile transition [63]. Also in soybean, 81 QTLs related to herbivory resistance were identified through meta-analysis, and the locations of true QTLs were deduced with a confidence interval of 95% [64, 65]. Thus, could it be determined whether a genetic variant having a particular QTL or haplotype of a polymorphism is associated with the resistant traits? To understand this contention, a number of herbivory-resistance QTLs have been tested for nonrandom associations in the populations derived from a cross between resistant and susceptible parents determining their proportional contribution to the phenotype in wide array of crop species [59, 66]. RFLP-based identification of herbivory-resistance QTLs in maize revealed their strong association with antixenosis (=a resistance mechanism employed by a plant to deter or prevent pest colonisation) and antibiosis (=an association of two organisms in which one is harmed or killed by the other) resistance to corn earworm [57, 67], and also in soybean [60]. The herbivory resistance QTLs discovered by Rector et al. [67] accounted for most of the genotypic variance for corn earworm resistance in the susceptible resistant hybrids; however, with some exceptions those could probably be addressed later with the help of soybean insect resistance QTL database. Will the contemporary catalogues of genome-sequencing projects across plant systems support the identification of important herbivory resistance loci? In result, it may definitely be assumed that future analyses based on whole-genome sequencing data will emphasize insect resistant (IR) QTLs/IR genes identified earlier through marker-assisted selection.

This will substantially reduce the time utilized for their adaptive inheritance through classical or precision breeding. Until relatively recently, family-based QTL mapping and association mapping were the primary means of searching genes involved in crop evolution [68]. Considering these studies, different chromosomal regions were identified harboring corresponding QTLs involved in the herbivory resistance phenotype. Because there is a complex correlation among different cellular traits considered important for a resistant phenotype, it becomes enormously difficult to identify such specific biochemical constituents. This observation suggests that variation in resistance traits is controlled by intricate genetic mechanisms, a suggestion further bolstered by demonstrations of resistance variation in maize synthetic hybrids, whose genomes have not undergone any subsequent selection [7, 17,69]. The mode of resistance is of great evolutionary interest, as it may often sporadically disappear under domestication and following breeding practices [7, 17,69]. So, what are such vital target genes, their chromosomal positions, and putative structural changes those cumulatively have influenced the loss of resistance potential? Will any type of stochastic mutations in the coding or noncoding regions lead to the differential loss of resistance potential among elite cultivars? From an evolutionary perspective, it could be hypothesized that divergence in resistance potential at the genomic level may also preserve an extensive polymorphism, thus retaining additional raw material for subsequent evolutionary tinkering if exploited under breeding programmes.

SUSCEPTIBILITY TO HERBIVORY AND ACQUIRED RESISTANCE

Resistance Genes: are There Patterns?

The domestication process increased a number of important traits required for agricultural innovations, though with few subsides. For example, the nascent "crop-form" (representative from independent domestication events) is more susceptible to herbivores as a result of having few important resistance loci, of which most were identified from their wild progenitor species. Is such phenotypic transformation under domestication universally advantageous or has accompanied with the loss of an "additional" benefit? What is the spectrum of consequences of having a set of important genomic loci selected under human selection? Are the answers to these questions consistent among plant lineages and/or between independent and parallel domestication events within a single species?

Traditional views maintain that domestication followed by breeding promoted the fixation of resistance loci, referred to as fixed heterozygosity [70–72]. It was thus suggested that inherited heterozygosity at resistance loci is beneficial. Modern views also support that introgression of resistance loci can be advantageous and provide a primary source of genes/alleles with new functions [73–75]. However, the identification of novel germplasm from the tertiary gene-pool is an enormously difficult task and it takes time for characterization. However, an inventive alternative is to carry out comprehensive genomic exploration of improved cultivars, primitive domesticated forms, and their wild progenitor species for the identification of candidate genes underlying resistance traits that show evidence of selection during domestication. With the latter approach, could the relationship between identified candidate genes with their phenotypic effects be envisaged? What are the confidence limits to forbid the possibility of the identified genes as false positives? The use of multiple statistical tests can certainly reduce such misreadings [76].

Analysis of a large number of loci underlying resistance to herbivory in soybean showed that resistance is an outcome of a mixture of major and minor gene effects and is not random. Some loci responsible for acquiring resistance to herbivory were underlying within regions having loci for the resistance against cyst nematode [77]. Classification of different soybean genotypes showing broad resistance has suggested important loci contributing to active synthesis and accumulation of products to stop (=antixenosis), deter (=antibiosis), and/or administer (for which mechanism is not readily established) herbivory [55, 67]. Three QTLs influencing resistance to corn borer species in maize have also been identified as overlapping with other genes of small effect in regulating the resistance phenotype, indicating the presence of pleiotropism or linkage between genes affecting resistance and other agronomic traits [78]. Recently, one major and three minor QTLs in rice have been identified as showing resistance against green rice leafhopper along with defined microsatellites for marker-assisted selection [79]. However, at present, relatively little is understood about the temporal dynamics of resistance to herbivory in different crop plants, and this requires the study of multiple genomes with the empirical reality of long-term resistance to herbivory.

Modern Cultivars Are More Susceptible Than Their Wild Progenitor Species

Under classical models, plant resistance to herbivory and pathogens is proposed to be a primary phenotype mostly available in the wild ancestors. For instance, plant introductions (PIs) in soybean with low agronomic quality have been demonstrated to be resistant against number of defoliating insects [12, 13]. Such models agree with the theory predicting that domestication occurred by human-mediated exertion through artificial selection on a wild species, both positive and negative, over hundreds of generations resulting in the development of cultivable species. In general, wild plant forms resist attack by herbivores and pathogens mainly through constitutive and inducible defense mechanisms [80]. The evolution and maintenance of the latter are now firmly accepted as an integral component of the plant defense mechanism against herbivores. However, the question remains when, where, and how induced resistance is deployed? Based on differences in the signaling pathways and spectra of effectiveness, the induced resistance could be categorized into (i) systemic acquired resistance (SAR) occurring in the distal plant parts following localized infections and (ii) induced systemic resistance (ISR) stimulated by nonpathogenic organisms and is regulated by jasmonic acid and ethylene [81]. During crop evolution (domestication followed by breeding practices), besides all evolutionarily relevant "internal" costs (genetic or allocation) of induced resistance, what are the other costs that may also be influencing the resistance phenotypes? There has been rapid progress in the detection of other important components, such as ecological costs, which are the result of a plant's interaction with its environment. Therefore, the conceptual separation of genetic and environmental contributions throughout crop formation would help in our understanding of induced resistance [82, 83].

If artificial selection prevails all through generations at the genomic level, is the operation of such selection global or localized? It may be argued that the evolutionary event such as domestication can affect the sequence variation at wide-reaching loci within a crop plant. In that scenario, what could be the limiting factors responsible for such proposed "genetic erosion" in an elite germplasm? One explanation could be that selection in modern breeding programs instead acts on selected important loci controlling a variety of traits, concluding that selection in either case would significantly reduce species-wide polymorphism and make it more vulnerable to the stressful conditions [4]. Besides intensive selection in modern breeding programs, the narrow genetic base is often cited as a contributing factor to low diversity, at least in soybean. Analysis of 111 fragments from 102 genes in four soybean populations showed evidence of a reduction in genetic diversity within and around the selected loci creating genetic bottlenecks [3]. The reduction of genetic diversity at different loci could result in "broad susceptibility" to newly emerging diseases and herbivores, thereby threatening long-term food and feed security [4]. One or multiple domestication events in the evolutionary history of soybean provide a discernible degree of diversity compensation that is up to a 50% reduction, eliminating almost 81% rare alleles present in the wild soybean (G. soja), and even appear to undergo significant change in the allele frequency [3]. Could it be assumed that the most significant loss of diversity in modern cultivars occurred during domestication, or due to an unusually low level of initial genetic variability in the wild progenitor, or both? It seems that the major loss of diversity occurred during domestication leading to the bottleneck where there was a loss of rare alleles present in the wild form and landraces. Contrary to classical predictions that loci under selection pressure may be relatively free to acquire heritable changes, it has been shown among subspecies of maize and rice that single nucleotide polymorphisms encode radical changes in the regulatory regions preferentially preserved in the domesticated forms and evolve conservatively [7, 17, 57]. Any such change can lead to divergence in subgenomic expression components, for example, in allopolyploid crop plants [53]. Those in result may influence the quantitatively inherited traits such as resistance to herbivory. Since wild plant forms have large genetic diversity that could be exploited for introgression of important traits in the modern varieties, it may be assumed that some sources of resistance have been left behind during plant domestication. However, it would be relatively difficult or even unsuccessful to introgress such traits into modern cultivars because this may increase the potential of inferior yield through linkage drag as also shown earlier [84, 85]. Moreover, it will also be difficult because crops with antifeedants from wild may have toxicity and anti-nutritional angle. Here the author has a strong opinion that transgenic technologies may provide a radical solution to the herbivory as these avoid linkage drag and also the antifeedant angle.

TRANSGENIC RESISTANCE MEDIATED BY THE EXPRESSION OF FOREIGN PROTEINS

Transgenic Crops and Resistance to Herbivory

Introduction of novel foreign genes into crop plants helps breeders to extend their germplasm with novel phenotypic traits. Such required traits are often related to the control of abiotic and biotic stresses, increasing the crop yield and improving the product quality, which were hitherto difficult or not possible to breed using a conventional approach [86, 87]. Given the toxicity of chemical pesticides, for the past two decades, a major emphasis has been on the control of herbivory through more rational strategies such as Integrated Pest Management (IPM). A component of IPM is the use of naturally available pesticides such as plant secondary metabolites and expression of heterologous proteins. Transgenic crops with a "modified" single gene developed for herbivore resistance are immensely beneficial in economic, environmental, and health concerns, as recently reviewed [88, 89], and understood to be "second generation" resistant crops (detailed in next section). A number of genes have been discovered which are toxic or antifeedant to herbivores and could be of plant or bacterial origin. A major contribution of herbivory-proof crops is in the reduction in application of harmful insecticides sprays and subsequent increase in crop yield [90, 91]. For example, transgenic maize event MON863 developed using a wild type gene from bacterium Bacillus thuringiensis (Bt), resistant to corn rootworm, was first commercialized in the USA in 2003 [92] and successfully grown until recently. In such a scenario, how do we provide a global perspective of the status of biotech crops? The easiest way is to calculate the global adoption rates as a percentage of the global areas of principal crops (i.e., soybean, cotton, maize, and canola) in which biotechnology is utilized. Though, during last one decade, herbicide tolerance has consistently been the dominant trait, deployment of multiple genes for other traits such as resistance to herbivores is becoming increasingly important and most prevalent for sustainable agriculture. The best example of the dynamics of this very rapid adoption is the contemporary biotech maize for stacked traits [93]. However, at the field level evaluation, the best-studied resistance to the herbivory phenomenon in the crops is the Bt cotton [94], which has a documented reduction in pesticide application of more than 70% in the developing countries, utmost domesticated in India [91], though results have been very variable. In 2009, 5.6 million small and marginal resource farmers planted and benefited from ~8.4 mHa of Bt cotton, equivalent to 8.7% of the total area under cotton cultivation in India. The corresponding adoption rate of biotech cotton has also been increased globally from 15.5 mHa to 16.1 mHa in the year 2009 [93]. For soybean, the global hectarage of herbicide tolerant transgenics was 69.2 mHa, (up by 3.4 mHa in 2009), which leads this crop to be the largest GM crop grown worldwide [93]. These examples illustrate two major contributions of the biotech approach to plant breeding: (i) enlarging the gene pool by including novel genes that breeders could not access by crossing techniques and (ii) modifying the genes by recombinant DNA technologies to fine-tune transgene expression [95]. The latter needs more emphasis and attention to achieve success in transgenic technologies for improved traits in the crop plants.

What controls the level of a foreign gene expression in genetically modified plants? Any or all of the molecular mechanisms associated with cellular gene expression machinery could be involved. Such a definition encompasses an array of molecular mechanisms at the transcriptional level including DNA methylation, mRNA decay, and small RNA-mediated gene silencing or at translational level having protein misfolding, degradation or other modification, and nuclear/chromosomal context with respect to genomic location of transgene often referred to as "position effect." It is generally assumed that because genes of prokaryotic origin are expressed poorly in higher organisms, such as plants [96, 97], certain modifications in such genes are required in order to achieve optimal expression. This may include modification in the GC content (as plants are comparatively GC rich than bacteria particularly Bacillus spp.), and in codon usage [97], after the introduction of regulatory sequences and polyadenylation signals. This fact is well supported with a significant increase in the expression of a codon-modified Bt cry1Ac gene in comparison to the wild-type gene from bacterial origin. Subsequently, by modifying codon usage, removing polyadenylation sequences [96], and other modifications in Bt genes, a number of plants have been transformed against their target pests [98,99]. But is this true only for bacterial genes? Can it be assumed that plant-derived genes with insecticidal potential [100, 101] such as protease-inhibitors, alpha-amylase inhibitor, lectins, and hemilectins do not require any modification for their optimal efficiency prior to the delivery into other crop systems?

Under field conditions, reverse effects of transgenics were also recorded with (i) harmful effect on nontarget insects and (ii) target insect to develop resistance against insecticidal genes used. The first risk could be addressed by designing the synthetic genes to target the hypervariable regions of target insect genes, thereby avoiding their lethal effects on nontarget insect population. However, in the latter case, the potential hazard could only be equilibrated through achieving very high transgene expression in the transgenics through the system-specific codon usage modification of a gene, use of high strength constitutive promoters, "position effect-" based screening of large transgenic population, and following refuge strategy to delay the acquired resistance in the pest population. If transgenic technologies are so promising and successful over conventional breeding, what are the major constraints that delayed the worldwide cultivation of genetically modified (GM) crops? Here, one of the likely explanations may be the inappropriate resistance management strategies deployed so far for the commercially important crop plants. For rice, assessment of agricultural fields for productivity and health effects in China emphasized such issues and highlighted key concerns on policy implementation and resolution of trade barriers [102], which may also be true globally. Clearly, much remains to be learned about such issues, as how should GM crops for herbivory resistance be synchronized for sustainable agriculture?

Next Generation Herbivory-Resistant Crop Plants

Transgenic technologies are undoubtedly important for plant defense against stresses, and it has also been argued that they are useful for the incorporation of novel phenotypes into crop plants. Existing single gene biotech crop studies may be extrapolated to a hypothetical case where full coverage of all target herbivores and plant diseases would be available in a genetic stock. As mentioned previously, biotech maize in USA is the best example of the deployment of stacked multiple traits including Bt genes (one to control the European corn borer complex and the other to control rootworm) and herbicide tolerance (first commercialized in 2005) and continued to grow in 2009 [93]. However, hints regarding gene pyramiding, exclusively for herbivory resistance, have also been emerging rapidly in last decade. In rice (Oryza indica), incorporation of two Bt genes and one lectin gene showed the control of three major herbivores: rice leaf folder (Cnaphalocrocis medinalis), yellow stemborer (Scirpophaga incertulas), and the brown planthopper (Nilaparvata lugens), respectively [103]. This indicates that, in rice, the long-term effect of multiple gene expression is an apparent enhancement of the resistance phenotype established by the synergistic effects of transgenes. Also, two Bt gene-transgenic cotton showed enhanced protection against Helicoverpa zea in comparison to the single gene transgenics with any of the two genes studied [104]. Evidence from our research laboratory on cotton indicates that transgenic stock with two Bt genes targeting two different lepidopteran insects shows significant improvement in the resistance trait against individual insects than do the respective single gene transgenics (B. Chaudhary and D. Pental, unpublished data). This indicates that the gene combinations may have played a strong role in reserving a durable resistance phenotype. Even without knowing the comprehensive specific mechanism(s) involved, there is clearly some association between the observation of high levels of protection against herbivores and the presence of multiple genes.

However, for any particular trait such as resistance to herbivores, can only genes from similar origin be tagged for synergistic activity? Can it be assumed that genes from distant origin be grouped together for increased resistance phenotype? The answer lies with the gut anatomy of phytophagous herbivore, which has different binding sites for toxic proteins that determine the spectrum of different lethal protein(s) activity and severity and provide clues for the use of diverse toxic genes to introgress herbivory resistance phenotype. The use of lectins, for example, Galanthus nivalis agglutinin (GNA), with Bt genes is very promising for increased insect resistance, with the ability of GNA to serve as carrier protein for the delivery of insecticidal proteins (in most cases Bt toxins)[105].

What should be the major selection criteria for insecticidal proteins when used for gene pyramiding? This may entail (i) toxic activity against wide spectrum of herbivores and (ii) nonhomology between or among concurrent toxins used. A well-studied example in second-generation insect-resistant transgenic plants is the use of novel vegetative insecticidal proteins (VIPs), which are produced by Bacillus thuringiensis during its vegetative growth along with Bt crystal proteins. Unlike Bt crystal proteins known as -endotoxins, VIPs are not parasporal and are secreted from the bacterial cell during vegetative growth. The full-length toxin gets activated proteolytically to a core toxin by proteases in the lepidopteran gut juice [106, 107]. Since the mode of action, structure, and binding sites of VIPs are different from Bt toxins in the insect gut epithelium, their use as potential insecticidal proteins for gene staking is very promising. Transgenic cotton expressing two insecticidal proteins Vip3A and Cry1Ab is estimated to be highly effective against two cotton herbivores,Helicoverpa armigera and Heliothis virescens [108]. Hence, it is assumed that enhanced resistance could be achieved by using two or more effective analogous insecticidal proteins. Gene combination Bt cry1Ac and snowdrop lectin GNA were also tested in cotton and showed resistance against insect pests Heliothis armigeraand Aphis gossypii [109]. Also, the results from a comparison between single Bt toxin tobacco transgenics and two gene transgenics having the Bt gene and the cowpea trypsin inhibitor (CpTi) gene showed the dominance of two gene products with enhanced insecticidal efficacy against cotton bollworm (Helicoverpa armigera) [110]. Further, an evaluation of Bt-CpTi fusion protein was performed in Brassica oleracea to study the insecticidal effect of this fusion protein on cabbage worm, which showed high activity of trypsin inhibitor, and the overall strong resistance to the common cabbage worm [111]. The synergistic activity between unrelated genes seems very promising. Having distinct binding sites of different Bt genes in the mid-gut of herbivores, novel combinations of such genes have been considered good to be deployed in delaying the evolution of resistant herbivores [112]. Since most of the activated Bt toxins share a common three-domain structure with a similar mode of action [113, 114], it is possible to develop a hybrid toxin through domain swapping [115, 116]. Transgenic cotton and tobacco with a hybrid Cry1EC toxin developed against polyphagus insect Spodoptera litura resulted in extreme toxicity to all developmental stages of larval development [116]. Naimov et al. [117] constructed a hybrid Bt gene using truncated cry1Ba gene as a scaffold and inserted part of the second domain of cryIIa gene conferring resistance to both coleopteran and lepidopteran pests. These studies support the assumption that such a novel strategy may provide new avenues for resistance management studies involving multiple transgenes in crop plants [88].

The pyramiding technology has been noted to provide excellent control of a broad range of herbivores and reinforces the argument that developed resistance in selected herbivores against one toxin will still be fully susceptible to other toxin molecules present in the plants (reviewed by [118]). If so, is gene pyramiding an enduring strategy for sustainable resistance? Given the insecticidal activity, it appears very clear that single gene transgenics in the field cannot be sustained without an integrated approach [119]. Such approach will definitely delay or, with other components of IPM strategies, preclude the possible adaptation of herbivore-populations to resistant transgenic plants.

Endogenous Resistance to Herbivores is prolonged by Transgene Stacking

As mentioned above, one strategy to delay the evolution of resistant herbivores is the stacking of multiple Btgenes [120, 121]. However, major concerns with this approach are (i) limited insecticidal properties of Btgenes to the target herbivores due to high specificity of Bt toxins, (ii) the potential cross-resistance leading to the evolution of resistant herbivores and (iii) the restricted use of possible novel combinations of Bt genes. A possible solution to the aforementioned problems may be either pyramiding of Bt genes with another transgene having a different mode of action, as also discussed earlier [108–110], or by pyramiding the native herbivory resistance genes with selected Bt transgenes (reviewed in [64]). In reference to the latter, Walker et al. [55]

identified that when the SSR-based IR-QTL conditioning corn earworm resistance in soybean was combined with Btcry1Ac gene, it resulted in detrimental effects on the larval weights, and with the least foliage consumption. With these results, can it be assumed that the combined effects of endogenous resistance and transgene-based resistance are additive while having independent mode of action(s)? Genetic modification of cotton with the Btcry1Ab transgene with high terpenoid levels showed more resistance to tobacco budworm than transgenics with low terpenoid levels [122]. However, transgenics developed in susceptible potato withBtcry3A gene against Colorado potato beetle larvae exhibit higher or at least similar mortality in the target herbivores as in the resistant potato line with leptine glycoalkaloids [123]. In the latter scenario, a better strategy may be the identification of IR-QTLs with the help of molecular markers, rather than with specific traits or compounds known to be associated with the resistance phenotype, as was earlier exampled in soybean, cotton, and potato [55, 122, 123]. The native plant resistance is suggested to be advantageous also in the controlling of the resistant herbivore populations, as demonstrated in the case of tobacco budworm, providing the "lethal dose" required for resistance management strategy [124]. Thus, staking multiple Btgenes along with the exploitation of native resistance through marker-assisted breeding will comprise complementary additive effects ameliorating the deployment of resistance management practices in the field.

CONCLUSIONS

What are the key evolutionary attributes that make the conversion of wild germplasm to "crop" such a prevalent phenomenon? Are the most important evolutionary properties of modern crop plants due to contemporary breeding efforts per se, or is domestication, artificial selection, just as important? In the opinion of a crop breeder, the end product is warranted by both. Plant breeding of independently selected domesticated forms began almost ten thousand years ago. Gene mutations occurred during selection, polyploidy, or artificial or natural hybridization and brought remarkable genetic variations. For management of stress conditions in crops, do parallel breeding efforts share similar genomic modifications or is it system/event specific, given the likelihood that different features of human-mediated selection may predominate among various lineages?

Very important successes during domestication in terms of crop yield and quality with other agronomic aspects have been achieved, but with compromised resistance to the herbivores and diseases. All modern crop plants are protected against herbivores by using synthetic toxic chemicals, as has been the case for many years, and this often leads to the development of resistance in herbivores against such frequently used chemicals. To circumvent this problem, introgression of important traits from the wild gene pool has been performed through classical breeding, but has met limited success due to incompatibility and numerous genetic and genomic differences among plant forms. Such principal distinctions may be critical, in that the interactions established by the initial conditions propagate from the time of initial origin through periods of stabilization and long-term evolutionary outcome.

Based on the limited number of available examples, when gene transfer occurred in the nascent interspecific crosses, genetic engineering techniques had proven useful to overcome such problems. Two potential benefits of genetic modification are precise transfer of foreign gene sequences and the evolution of adaptive transgressive traits. Two potential benefits of genetic modification are precise transfer of foreign gene sequences and the evolution of adaptive transgressive traits. Classic plant breeding programs are reinstated for heterozygosity and therefore may be more likely to experience locus-specific linkage with its evolutionary consequences. In contrast, a transgenic crop plant possesses a greater insertion of multiple alien genic sequences with immediate phenotypic effect and genetic novelty. Indeed, experimental analysis of genetically modified crops for multiple traits suggests that pyramiding of favorable genes for individual trait may yield a "super-crop" with high returns. Thus, the breadth of recurrently selected traits in the domesticated plants and the genetic transformation system together has a major effect on the creation and retention of evolutionary novelty in the crop system.

ACKNOWLEDGMENTS

The author is thankful to Professor Jonathan F. Wendel and Kara Grupp, Iowa State University, USA, for their suggestions in the preparation of this paper. The author also thanks the Department of Science and

Technology (DST), Council of Scientific and Industrial Research (CSIR), and the Department of Biotechnology (DBT), Government of India, for the financial support to carry out cotton research work in the laboratory.

REFERENCES

1. J. C. Burger, M. A. Chapman, and J. M. Burke, "Molecular insights into the evolution of crop plants,"American Journal of Botany, vol. 95, pp. 113–122, 2008.
2. C. Darwin, On the Origin of Species by Means of Natural Selection, or the Preservation of Favoured Races in the Struggle for Life, W. Clowes and Sons and Charing Cross, London, UK, 1859.
3. D. L. Hyten, Q. Song, Y. Zhu et al., "Impacts of genetic bottlenecks on soybean genome diversity,"Proceedings of the National Academy of Sciences of the United States of America, vol. 103, no. 45, pp. 16666–16671, 2006.
4. S. D. Tanksley and S. R. McCouch, "Seed banks and molecular maps: unlocking genetic potential from the wild," Science, vol. 277, no. 5329, pp. 1063–1066, 1997.
5. J. J. Bull and H. A. Wichmann, "Applied evolution," Annual Review of Ecology, Evolution, and Systematics, vol. 32, pp. 183–217, 2001.
6. T. D. Clarkson, "Stress physiology in crop plants," Plant, Cell & Environment, vol. 4, p. 184, 1981.
7. J. Doebley, "The genetics of maize evolution," Annual Review of Genetics, vol. 38, pp. 37–59, 2004.
8. J. Doebley, A. Stec, and L. Hubbard, "The evolution of apical dominance in maize," Nature, vol. 386, no. 6624, pp. 485–488, 1997.
9. T. Komatsuda, M. Pourkheirandish, C. He et al., "Six-rowed barley originated from a mutation in a homeodomain-leucine zipper I-class homeobox gene," Proceedings of the National Academy of Sciences of the United States of America, vol. 104, no. 4, pp. 1424–1429, 2007.
10. C. Li, A. Zhou, and T. Sang, "Rice domestication by reducing shattering," Science, vol. 311, no. 5769, pp. 1936–1939, 2006.

11. M. T. Sweeney, M. J. Thomson, B. E. Pfeil, and S. McCouch, "Caught red-handed: Rc encodes a basic helix-loop-helix protein conditioning red pericarp in rice," Plant Cell, vol. 18, no. 2, pp. 283–294, 2006
12. J. N. All, H. R. Boerma, and J. W. Todd, "Screening soybean genotypes in the greenhouse for resistance to insects," Crop Science, vol. 29, pp. 1156–1159, 1989.
13. V. D. Luedders and W. A. Dickerson, "Resistance of selected soybean genotypes and segregating populations to cabbage looper feeding," Crop Science, vol. 17, pp. 395–396, 1977.
14. T. Ashfield, J. R. Danzer, D. Held et al., "Rpg1, a soybean gene effective against races of bacterial blight, maps to a cluster of previously identified disease resistance genes," Theoretical and Applied Genetics, vol. 96, no. 8, pp. 1013–1021, 1998.
15. A. Garcia, É. S. Calvo, R. A. De Souza Kiihl, A. Harada, D. M. Hiromoto, and L. G. E. Vieira, "Molecular mapping of soybean rust (Phakopsora pachyrhizi) resistance genes: discovery of a novel locus and alleles," Theoretical and Applied Genetics, vol. 117, no. 4, pp. 545–553, 2008.
16. B. Chaudhary, R. Hovav, R. Rapp, N. Verma, J. A. Udall, and J. F. Wendel, "Global analysis of gene expression in cotton fibers from wild and domesticated Gossypium barbadense," Evolution and Development, vol. 10, no. 5, pp. 567–582, 2008.
17. J. Doebley, "Unfallen grains: how ancient farmers turned weeds into crops," Science, vol. 312, no. 5778, pp. 1318–1319, 2006.
18. M. Sweeney and S. McCouch, "The complex history of the domestication of rice," Annals Botany, vol. 100, pp. 951–957, 2007.
19. N. I. Vavilov, "The origin, variation, immunity and breeding of cultivated plants," Chronica Botanica, vol. 13, no. 1–6, pp. 1–366, 1951.
20. P. Gepts, "A comparison between crop domestication, classical plant breeding, and genetic engineering," Crop Science, vol. 42, no. 6, pp. 1780–1790, 2002.
21. H. Ulukan, "The evolution of cultivated plant species: classical plant breeding versus genetic engineering," Plant Systematics and Evolution, vol. 280, no. 3-4, pp. 133–142, 2009.

22. H. W. Cai and H. Morishima, "QTL clusters reflect character associations in wild and cultivated rice,"Theoretical and Applied Genetics, vol. 104, no. 8, pp. 1217–1228, 2002.
23. L. Z. Xiong, K. D. Liu, X. K. Dai, C. G. Xu, and Q. Zhang, "Identification of genetic factors controlling domestication-related traits of rice using an F2 population of a cross between Oryza sativa and O. rufipogon," Theoretical and Applied Genetics, vol. 98, no. 2, pp. 243–251, 1999.
24. T. Isemura, A. Kaga, S. Konishi et al., "Genome dissection of traits related to domestication in azuki bean (Vigna angularis) and comparison with other warm-season legumes," Annals of Botany, vol. 100, no. 5, pp. 1053–1071, 2007.
25. R. A. Rapp, C. H. Haigler, L. Flagel, R. H. Hovav, J. A. Udall, and J. F. Wendel, "Gene expression in developing fibres of Upland cotton (Gossypium hirsutum L.) was massively altered by domestication,"BMC Biology, vol. 8, article 139, 2010.
26. Y. Bai and P. Lindhout, "Domestication and breeding of tomatoes: what have we gained and what can we gain in the future?" Annals of Botany, vol. 100, no. 5, pp. 1085–1094, 2007.
27. M. C. Simonetti, M. P. Bellomo, G. Laghetti, P. Perrino, R. Simeone, and A. Blanco, "Quantitative trait loci influencing free-threshing habit in tetraploid wheats," Genetic Resources and Crop Evolution, vol. 46, no. 3, pp. 267–271, 1999.
28. V. J. Nalam, M. I. Vales, C. J. W. Watson, E. B. Johnson, and O. Riera-Lizarazu, "Map-based analysis of genetic loci on chromosome 2D that affect glume tenacity and threshability, components of the free-threshing habit in common wheat (Triticum aestivum L.)," Theoretical and Applied Genetics, vol. 116, no. 1, pp. 135–145, 2007.
29. N. Watanabe, Y. Fujii, N. Kato, T. Ban, and P. Martinek, "Microsatellite mapping of the genes for brittle rachis on homoeologous group 3 chromosomes in tetraploid and hexaploid wheats," Journal of Applied Genetics, vol. 47, no. 2, pp. 93–98, 2006.
30. D. Zohary and M. Hopf, Domestication of Plants in the Old World, Oxford University Press, Oxford, UK, 3rd edition, 2000.

31. S. A. Kempin, B. Savidge, and M. F. Yanofsky, "Molecular basis of the cauliflower phenotype in arabidopsis," Science, vol. 267, no. 5197, pp. 522–525, 1995.
32. G. Ladizinsky, "Lentil domestication: on the quality of evidence and arguments," Economic Botany, vol. 47, no. 1, pp. 60–64, 1993.
33. N. F. Weeden, "Genetic changes accompanying the domestication of Pisum sativum: is there a common genetic basis to the "domestication syndrome" for legumes?" Annals of Botany, vol. 100, no. 5, pp. 1017–1025, 2007.
34. D. Ugent, "The potato," Science, vol. 170, no. 3963, pp. 1161–1166, 1970.
35. B. D. Smith, "The initial domestication of Cucurbita pepo in the Americas 10,000 years ago," Science, vol. 276, no. 5314, pp. 932–934, 1997.
36. T. Sang, "Genes and mutations underlying domestication transitions in grasses," Plant Physiology, vol. 149, no. 1, pp. 63–70, 2009.
37. A. H. Paterson, J. F. Wendel, H. Gundlach et al., "Repeated polyploidization of Gossypium genomes and the evolution of spinnable cotton fibres," Nature, vol. 492, pp. 423–427, 2012.
38. J. Dubcovsky and J. Dvorak, "Genome plasticity a key factor in the success of polyploid wheat under domestication," Science, vol. 316, no. 5833, pp. 1862–1866, 2007.
39. W. Li and B. S. Gill, "Multiple genetic pathways for seed shattering in the grasses," Functional and Integrative Genomics, vol. 6, no. 4, pp. 300–309, 2006.
40. J. C. Reif, P. Zhang, S. Dreisigacker et al., "Wheat genetic diversity trends during domestication and breeding," Theoretical and Applied Genetics, vol. 110, no. 5, pp. 859–864, 2005.
41. S. I. Wright, I. V. Bi, S. C. Schroeder et al., "Evolution: the effects of artificial selection on the maize genome," Science, vol. 308, no. 5726, pp. 1310–1314, 2005.
42. J. Doebley and L. Lukens, "Transcriptional regulators and the evolution of plant form," Plant Cell, vol. 10, no. 7, pp. 1075–1082, 1998.

43. J. F. Doebley, B. S. Gaut, and B. D. Smith, "The molecular genetics of crop domestication," Cell, vol. 127, no. 7, pp. 1309–1321, 2006.

44. S. R. Palumbi, "Humans as the world's greatest evolutionary force," Science, vol. 293, pp. 1786–1790, 2001.

45. M. P. Bretó, M. J. Asins, and E. A. Carbonell, "Genetic variability in Lycopersicon species and their genetic relationships," Theoretical and Applied Genetics, vol. 86, no. 1, pp. 113–120, 1993.

46. H. Egashira, H. Ishihara, T. Takashina, and S. Imanishi, "Genetic diversity of the "peruvianum-complex" (Lycopersicon peruvianum (L.) Mill. and L. chilense Dun.) Revealed by RAPD analysis,"Euphytica, vol. 116, no. 1, pp. 23–31, 2000.

47. J. Villand, P. W. Skroch, T. Lai, P. Hanson, C. G. Kuo, and J. Nienhuis, "Genetic variation among tomato accessions from primary and secondary centers of diversity," Crop Science, vol. 38, no. 5, pp. 1339–1347, 1998.

48. C. M. Rick and R. T. Chetelat, "Utilization of related wild species for tomato improvement," Acta Horticulturae, vol. 412, pp. 21–38, 1995.

49. J. C. Miller and S. D. Tanksley, "RFLP analysis of phylogenetic relationships and genetic variation in the genus Lycopersicon," Theoretical and Applied Genetics, vol. 80, no. 4, pp. 437–448, 1990.

50. S. García-Martínez, L. Andreani, M. Garcia-Gusano, F. Geuna, and J. J. Ruiz, "Evaluation of amplified fragment length polymorphism and simple sequence repeats for tomato germplasm fingerprinting: utility for grouping closely related traditional cultivars," Genome, vol. 49, no. 6, pp. 648–656, 2006.

51. S. M. Tam, C. Mhiri, A. Vogelaar, M. Kerkveld, S. R. Pearce, and M. A. Grandbastien, "Comparative analyses of genetic diversities within tomato and pepper collections detected by retrotransposon-based SSAP, AFLP and SSR," Theoretical and Applied Genetics, vol. 110, no. 5, pp. 819–831, 2005.

52. J. F. Wendel and R. C. Cronn, "Polyploidy and the evolutionary history of cotton," Advances in Agronomy, vol. 78, pp. 139–186, 2003.

53. B. Chaudhary, L. Flagel, R. M. Stupar et al., "Reciprocal silencing, transcriptional bias and functional divergence of homeologs in polyploid cotton (Gossypium)," Genetics, vol. 182, no. 2, pp. 503–517, 2009.
54. S. J. O'Brien, Genetic Maps, Cold Spring Harbor Laboratory Press, Cold Spring Harbor, NY, USA, 6th edition, 1993.
55. D. R. Walker, J. M. Narvel, H. R. Boerma, J. N. All, and W. A. Parrott, "A QTL that enhances and broadens Bt insect resistance in soybean," Theoretical and Applied Genetics, vol. 109, no. 5, pp. 1051–1057, 2004.
56. J. G. K. Williams, A. R. Kubelik, K. J. Livak, J. A. Rafalski, and S. V. Tingey, "DNA polymorphisms amplified by arbitrary primers are useful as genetic markers," Nucleic Acids Research, vol. 18, no. 22, pp. 6531–6535, 1990.
57. P. F. Byrne, M. D. McMullen, B. R. Wiseman et al., "Identification of maize chromosome regions associated with antibiosis to corn earworm (Lepidoptera: Noctuidae) larvae," Journal of Economic Entomology, vol. 90, no. 4, pp. 1039–1045, 1997.
58. N. D. Young, L. Kumar, D. Menancio-Hautea et al., "RFLP mapping of a major bruchid resistance gene in mungbean (Vigna radiata, L. Wilczek)," Theoretical and Applied Genetics, vol. 84, no. 7-8, pp. 839–844, 1992.
59. G. C. Yencho, M. W. Bonierbale, W. M. Tingey, R. L. Plaisted, and S. D. Tanksley, "Molecular markers locate genes for resistance to the Colorado potato beetle, Leptinotarsa decemlineata, in hybrid Solanum tuberosum × S. berthaultii potato progenies," Entomologia Experimentalis et Applicata, vol. 81, no. 2, pp. 141–154, 1996.
60. J. M. Narvel, D. R. Walker, B. G. Rector, J. N. All, W. A. Parrott, and H. R. Boerma, "A retrospective DNA marker assessment of the development of insect resistant soybean," Crop Science, vol. 41, no. 6, pp. 1931–1939, 2001.
61. D. Kliebenstein, D. Pedersen, B. Barker, and T. Mitchell-Olds, "Comparative analysis of quantitative trait loci controlling glucosinolates, myrosinase and insect resistance in Arabidopsis thaliana," Genetics, vol. 161, no. 1, pp. 325–332, 2002.

62. T. D. Brooks, B. S. Bushman, W. P. Williams, M. D. McMullen, and P. M. Buckley, "Genetic basis of resistance to fall armyworm (Lepidoptera: Noctuidae) and southwestern corn borer (Lepidoptera: Crambidae) leaf-feeding damage in maize," Journal of Economic Entomology, vol. 100, no. 4, pp. 1470–1475, 2007.
63. W. P. Williams, F. M. Davis, P. M. Buckley, P. A. Hedin, G. T. Baker, and D. S. Luthe, "Factors associated with resistance to fall armyworm (Lepidoptera: Noctuidae) and Southwestern corn borer (Lepidoptera: Crambidae) in corn at different vegetative stages," Journal of Economic Entomology, vol. 91, no. 6, pp. 1471–1480, 1998.
64. H. R. Boerma and D. R. Walker, "Discovery and utilization of QTLs for insect resistance in soybean,"Genetica, vol. 123, no. 1-2, pp. 181–189, 2005.
65. J. Wang, W. K. Song, W. B. Zhang, C. Y. Liu, G. H. Hu, and Q. S. Chen, "Meta-analysis of insect-resistance QTLs in soybean," Yi Chuan, vol. 31, no. 9, pp. 953–961, 2009.
66. S. D. Tanksley, N. D. Young, A. H. Paterson, and M. W. Bonierbale, "RFLP mapping in plant breeding: new tools for an old science," BioTechnology, vol. 7, pp. 257–264, 1989.
67. B. G. Rector, J. N. All, W. A. Parrott, and H. R. Boerma, "Quantitative trait loci for antibiosis resistance to corn earworm in soybean," Crop Science, vol. 40, no. 1, pp. 233–238, 2000.
68. A. H. Paterson, "What has QTL mapping taught us about plant domestication?" New Phytologist, vol. 154, no. 3, pp. 591–608, 2002.
69. M. Guo, S. Yang, M. Rupe et al., "Genome-wide allele-specific expression analysis using Massively Parallel Signature Sequencing (MPSS) Reveals cis- and trans-effects on gene expression in maize hybrid meristem tissue," Plant Molecular Biology, vol. 66, no. 5, pp. 551–563, 2008.
70. A. L. Kahler, M. I. Morris, and R. W. Allard, "Gene triplication and fixed heterozygosity in diploid wild barley," Journal of Heredity, vol. 72, no. 6, pp. 374–376, 1981.
71. A. B. Schooler, "Wild barley hybrids," Journal of Heredity, vol. 54, no. 3, pp. 130–132, 1963.

72. C. A. Suneson, "Genetic diversity—a protection against plant diseases and insects," Agronomy Journal, vol. 52, pp. 319–321, 1960.
73. R. W. Michelmore and B. C. Meyers, "Clusters of resistance genes in plants evolve by divergent selection and a birth-and-death process," Genome Research, vol. 8, no. 11, pp. 1113–1130, 1998.
74. D. P. Singh and A. Singh, Disease and Insect Resistance in Plants, Science Publisher, Enfield, NH, USA, 2005.
75. K. Zhao, M. Wright, J. Kimball et al., "Genomic diversity and introgression in O. sativa reveal the impact of domestication and breeding on the rice genome," PloS ONE, vol. 5, no. 5, Article ID e10780, 2010.
76. A. Bonin, P. Taberlet, C. Miaud, and F. Pompanon, "Explorative genome scan to detect candidate loci for adaptation along a gradient of altitude in the common frog (Rana temporaria)," Molecular Biology and Evolution, vol. 23, no. 4, pp. 773–783, 2006.
77. C. R. Yesudas, H. Sharma, and D. A. Lightfoot, "Identification of QTL in soybean underlying resistance to herbivory by Japanese beetles (Popillia japonica, Newman)," Theoretical and Applied Genetics, vol. 121, no. 2, pp. 353–362, 2010.
78. B. Ordas, R. A. Malvar, R. Santiago, and A. Butron, "QTL mapping for Mediterranean corn borer resistance in European flint germplasm using recombinant inbred lines," BMC Genomics, vol. 11, no. 1, article 174, 2010.
79. D. Fujita, K. Doi, A. Yoshimura, and H. Yasui, "A major QTL for resistance to green rice leafhopper (Nephotettix cincticeps Uhler) derived from African rice (Oryza glaberrima Steud.)," Breeding Science, vol. 60, no. 4, pp. 336–341, 2010.
80. D. Cipollini and M. Heil, "Costs and benefits of induced resistance to herbivores and pathogens in plants," in CAB Reviews: Perspectives in Agriculture, Veterinary Science, Nutrition and Natural Resources, p. 10, 2010.
81. D. Walters and M. Heil, "Costs and trade-offs associated with induced resistance," Physiological and Molecular Plant Pathology, vol. 71, no. 1–3, pp. 3–17, 2007.

82. M. Heil, "Ecological costs of induced resistance," Current Opinion in Plant Biology, vol. 5, no. 4, pp. 345–350, 2002.
83. M. Heil and I. T. Baldwin, "Fitness costs of induced resistance: emerging experimental support for a slippery concept," Trends in Plant Science, vol. 7, no. 2, pp. 61–67, 2002.
84. N. D. Young and S. D. Tanksley, "RFLP analysis of the size of chromosomal segments retained around the Tm-2 locus of tomato during backcross breeding," Theoretical and Applied Genetics, vol. 77, no. 3, pp. 353–359, 1989.
85. A. C. Zeven, D. R. Knott, and R. Johnson, "Investigation of linkage drag in near isogenic lines of wheat by testing for seedling reaction to races of stem rust, leaf rust and yellow rust," Euphytica, vol. 32, no. 2, pp. 319–327, 1983.
86. W. Schuch, J. M. Kanczler, D. Robertson et al., "Fruit quality characteristics of transgenic tomato fruit with altered polygalacturonase activity," HortScience, vol. 26, pp. 1517–1520, 1991.
87. S. H. Strauss, "Genomics, genetic engineering, and domestication of crops," Science, vol. 300, no. 5616, pp. 61–62, 2003.
88. P. Christou, T. Capell, A. Kohli, J. A. Gatehouse, and A. M. R. Gatehouse, "Recent developments and future prospects in insect pest control in transgenic crops," Trends in Plant Science, vol. 11, no. 6, pp. 302–308, 2006.
89. N. Ferry, M. G. Edwards, J. Gatehouse, T. Capell, P. Christou, and A. M. R. Gatehouse, "Transgenic plants for insect pest control: a forward looking scientific perspective," Transgenic Research, vol. 15, no. 1, pp. 13–19, 2006.
90. S. M. High, M. B. Cohen, Q. Y. Shu, and I. Altosaar, "Achieving successful deployment of Bt rice,"Trends in Plant Science, vol. 9, no. 6, pp. 286–292, 2004.
91. M. Qaim and D. Zilberman, "Yield effects of genetically modified crops in developing countries,"Science, vol. 299, no. 5608, pp. 900–902, 2003.
92. T. Vaughn, T. Cavato, G. Brar et al., "A method of controlling corn rootworm feeding using a Bacillus thuringiensis protein expressed in transgenic maize," Crop Science, vol. 45, no. 3, pp. 931–938, 2005.

93. C. James, "Global status of commercialized biotech/GM crops: 2009," ISAAA Brief 41 ISAAA,http://www.isaaa.org.

94. F. J. Perlak, R. W. Deaton, T. A. Armstrong et al., "Insect resistant cotton plants," Bio/Technology, vol. 8, no. 10, pp. 939–943, 1990.

95. C. N. Stewart, M. J. Adang, J. N. All et al., "Genetic transformation, recovery, and characterization of fertile soybean transgenic for a synthetic Bacillus thuringiensis cryIAc gene," Plant Physiology, vol. 112, no. 1, pp. 121–129, 1996.

96. S. H. Diehn, W. L. Chiu, E. Jay De Rocher, and P. J. Green, "Premature polyadenylation at multiple sites within a Bacillus thuringiensis toxin gene-coding region," Plant Physiology, vol. 117, no. 4, pp. 1433–1443, 1998.

97. F. J. Perlak, R. L. Fuchs, D. A. Dean, S. L. McPherson, and D. A. Fischhoff, "Modification of the coding sequence enhances plant expression of insect control protein genes," Proceedings of the National Academy of Sciences of the United States of America, vol. 88, no. 8, pp. 3324–3328, 1991.

98. B. Chaudhary, Development of transgenic lines in cotton (Gossypium hirsutum L. cv. Coker 310FR) for insect resistance and marker gene removal [Ph.D. thesis], Department of Genetics, University of Delhi South Campus, New Delhi, India, 2006.

99. B. Chaudhary, S. Kumar, K. V. S. K. Prasad, G. S. Oinam, P. K. Burma, and D. Pental, "Slow desiccation leads to high-frequency shoot recovery from transformed somatic embryos of cotton (Gossypium hirsutum L. cv. Coker 310 FR)," Plant Cell Reports, vol. 21, no. 10, pp. 955–960, 2003.

100. Y. E. Dunaevsky, E. N. Elpidina, K. S. Vinokurov, and M. A. Belozersky, "Protease inhibitors in improvement of plant resistance to pathogens and insects," Molecular Biology, vol. 39, no. 4, pp. 608–613, 2005.

101. L. L. Murdock and R. E. Shade, "Lectins and protease inhibitors as plant defenses against insects,"Journal of Agricultural and Food Chemistry, vol. 50, no. 22, pp. 6605–6611, 2002.

102. J. Huang, R. Hu, S. Rozelle, and C. Pray, "Plant science: insect-resistant GM rice in farmers' fields: assessing productivity and health effects in China," Science, vol. 308, no. 5722, pp. 688–690, 2005

103. S. B. Maqbool, S. Riazuddin, N. T. Loc, A. M. R. Gatehouse, J. A. Gatehouse, and P. Christou, "Expression of multiple insecticidal genes confers broad resistance against a range of different rice pests," Molecular Breeding, vol. 7, no. 1, pp. 85–93, 2001.

104. R. E. Jackson, J. R. Bradley, and J. W. Van Duyn, "Performance of feral and Cry1Ac-selected Helicoverpa zea (Lepidoptera: Noctuidae) strains on transgenic cottons expressing one or two Bacillus thuringiensisssp. kurstaki proteins under greenhouse conditions," Journal of Entomological Science, vol. 39, no. 1, pp. 46–55, 2004.

105. E. Fitches, M. G. Edwards, C. Mee et al., "Fusion proteins containing insect-specific toxins as pest control agents: snowdrop lectin delivers fused insecticidal spider venom toxin to insect haemolymph following oral ingestion," Journal of Insect Physiology, vol. 50, no. 1, pp. 61–71, 2004.

106. R. A. De Maagd, A. Bravo, C. Berry, N. Crickmore, and H. E. Schnepf, "Structure, diversity, and evolution of protein toxins from spore-forming entomopathogenic bacteria," Annual Review of Genetics, vol. 37, pp. 409–433, 2003.

107. C. G. Yu, M. A. Mullins, G. W. Warren, M. G. Koziel, and J. J. Estruch, "The Bacillus thuringiensisvegetative insecticidal protein Vip3A lyses midgut epithelium cells of susceptible insects," Applied and Environmental Microbiology, vol. 63, no. 2, pp. 532–536, 1997.

108. R. W. Kurtz, A. McCaffery, and D. O'Reilly, "Insect resistance management for Syngenta's VipCot transgenic cotton," Journal of Invertebrate Pathology, vol. 95, no. 3, pp. 227–230, 2007.

109. S. H. Syed, Genetic transformation of cotton with galanthus nivalis agglutinin (GNA) gene [Ph.D. thesis], University of the Punjab, Lahore, Pakistan, 2002.

110. X. Fan, X. Shi, J. Zhao, R. Zhao, and Y. Fan, "Insecticidal activity of transgenic tobacco plants expressing both Bt and CpTI genes on cotton bollworm (Helicoverpa armigera)," Chinese Journal of Biotechnology, vol. 15, no. 1, pp. 1–5, 1999.

111. G. D. Yang, Z. Zhu, Y. O. Li, and Z. J. Zhu, "Transformation of Bt-CpTi fusion protein gene to cabbage (Brassica oleracea var. capitata) mediated by Agrobacterium tumefaciens and particle bombardment,"Shi Yan Sheng Wu Xue Bao, vol. 35, no. 2, pp. 117–122, 2002.

112. R. E. Jackson, J. R. Bradley, and J. W. Van Duyn, "Field performance of transgenic cottons expressing one or two Bacillus thuringiensis endotoxins against bollworm, Helicoverpa zea (Boddie)," Journal of Cotton Science, vol. 7, no. 3, pp. 57–64, 2003.
113. R. A. De Maagd, A. Bravo, and N. Crickmore, "How Bacillus thuringiensis has evolved specific toxins to colonize the insect world," Trends in Genetics, vol. 17, no. 4, pp. 193–199, 2001.
114. E. Schnepf, N. Crickmore, J. Van Rie et al., "Bacillus thuringiensis and its pesticidal crystal proteins,"Microbiology and Molecular Biology Reviews, vol. 62, no. 3, pp. 775–806, 1998.
115. R. Karlova, M. Weemen-Hendriks, S. Naimov, J. Ceron, S. Dukiandjiev, and R. A. De Maagd, "Bacillus thuringiensis - endotoxin Cry1Ac domain III enhances activity against Heliothis virescens in some, but not all Cry1-Cry1Ac hybrids," Journal of Invertebrate Pathology, vol. 88, no. 2, pp. 169–172, 2005.
116. P. Singh, M. Kumar, C. Chaturvedi, D. Yadav, and R. Tuli, "Development of a hybrid -endotoxin and its expression in tobacco and cotton for control of a polyphagous pest Spodoptera litura," Transgenic Research, vol. 13, no. 5, pp. 397–410, 2004.
117. S. Naimov, S. Dukiandjiev, and R. de Maagd, "A hybrid Bacillus thuringiensis delta-endotoxin gives resistance against a coleopteran and a lepidopteran pest in transgenic potato," Plant Biotechnology, vol. 1, pp. 51–57, 2003.
118. W. Manyangarirwa, M. Turnbull, G. S. McCutcheon, and J. P. Smith, "Gene pyramiding as a Btresistance management strategy: how sustainable is this strategy?" African Journal of Biotechnology, vol. 5, no. 10, pp. 781–785, 2006.
119. S. L. Bates, J. Z. Zhao, R. T. Roush, and A. M. Shelton, "Insect resistance management in GM crops: past, present and future," Nature Biotechnology, vol. 23, no. 1, pp. 57–62, 2005.
120. R. T. Roush, "Two-toxin strategies for management of insecticidal transgenic crops: can pyramiding succeed where pesticide mixtures have not?" Philosophical Transactions of the Royal Society B, vol. 353, no. 1376, pp. 1777–1786, 1998.
121. B. E. Tabashnik, "Evolution of resistance to Bacillus thuringiensis," Annual Review of Entomology, vol. 39, pp. 47–79, 1994.

122. E. S. Sachs, J. H. Benedict, J. F. Taylor, D. M. Stelly, S. K. Davis, and D. W. Altman, "Pyramiding CryIA(b) insecticidal protein and terpenoids in cotton to resist tobacco budworm (Lepidoptera: Noctuidae)," Environmental Entomology, vol. 25, no. 6, pp. 1257–1266, 1996.
123. J. J. Coombs, D. S. Douches, W. Li, E. J. Grafius, and W. L. Pett, "Combining engineered (Bt-cry3A) and natural resistance mechanisms in potato for control of Colorado potato beetle," Journal of the American Society for Horticultural Science, vol. 127, no. 1, pp. 62–68, 2002.
124. D. Walker, H. Roger Boerma, J. All, and W. Parrott, "Combining Cry1Ac with QTL alleles from PI 229358 to improve soybean resistance to lepidopteran pests," Molecular Breeding, vol. 9, no. 1, pp. 43–51, 2002.

Chapter 4

Farmers without Borders-Genetic Structuring in Century Old Barley (Hordeum vulgare)

N E G Forsberg[1,2], J Russell[3], M Macaulay[3], M W Leino[2,4], and J Hagenblad[2]

[1]Norwegian University of Science and Technology, Department of Biology, Trondheim, Norway

[2]IFM-Biology, Linköping University, Linköping, Sweden

[3]The James Hutton Institute, Invergowrie, Dundee, Scotland, UK

[4]Swedish Museum of Cultural History, Julita, Sweden

ABSTRACT

The geographic distribution of genetic diversity can reveal the evolutionary history of a species. For crop plants, phylogeographic patterns also indicate how seed has been exchanged and spread in agrarian communities. Such patterns are, however, easily blurred by the intense seed trade, plant improvement and even genebank conservation during the twentieth century, and discerning fine-scale phylogeographic patterns is thus particularly challenging. Using historical crop specimens, these problems are circumvented and we show here how high-throughput genotyping of historical nineteenth century crop specimens can reveal detailed geographic population structure. Thirty-one historical and nine extant accessions of North European landrace barley (*Hordeum vulgare* L.), in total 231 individuals, were genotyped on a 384 single nucleotide polymorphism assay. The historical material shows constant high levels of within-accession diversity, whereas the extant accessions show more varying levels of diversity and a higher degree of total genotype sharing. *Structure,* discriminant analysis of principal components and principal component analysis cluster the accessions in latitudinal groups across country borders in Finland, Norway and Sweden. F_{ST} statistics indicate strong differentiation between accessions from southern Fennoscandia and accessions from central or northern Fennoscandia, and less differentiation between central and northern accessions. These findings are discussed in the context of contrasting historical records on intense within-country south to north seed movement. Our results suggest that although seeds were traded long distances, long-term cultivation has instead been of locally available, possibly better adapted, genotypes.

INTRODUCTION

Population genetics and phylogeography are important tools that provide insight into the evolutionary history of species. Geographic patterns in the distribution of genetic diversity can give information about the geographic origin of lineages or the effects of migration routes (Avise, 2009) and have, among other species, been applied to crop plants (for example, Olsson and Schaal, 1999; Londo *et al.*, 2006; Saisho and Purugganan, 2007). Insights into the genetic structure of crop species provide not only a better understanding of their evolutionary history but can also increase the knowledge about cultural exchange in agrarian communities (Van Heerwaarden *et al.*, 2011; Oliveira *et al.*, 2012; Roullier *et al.*, 2013).

Barley (*Hordeum vulgare* L. ssp. *vulgare*) was domesticated about 10 000 years ago (Badr *et al.*, 2000), most likely from multiple domestication centers (Morrell and Clegg, 2007; Saisho and Purugganan,

2007; Fuller *et al.*, 2011; Ren *et al.*, 2013). An adaptable species, barley has been of great importance also in regions where climate or soil is sub-optimal for agriculture. As a result of crop improvement during the twentieth century, present-day cultivars primarily show phylogeographic structuring on a continental scale. Agronomic traits, such as winter or spring growth habit and two-row or six-row type, have been found to be as relevant as geographic origin in determining population clustering (Malysheva-Otto *et al.*, 2006).

To detect traces of evolutionary history, landraces are a preferable choice to modern cultivars. The definition of a landrace is not without controversy (Zeven, 1998; Camacho Villa *et al.*, 2005), but they can be described as locally adapted populations, genetically diverse and with a historical origin, lacking formal crop improvement. During centuries of continuous cultivation in their respective area, their genetic composition has shifted because of gene flow and seed trade as well as local adaptation and genetic drift, but landraces are generally considered to have been relatively stable over time (Brown, 1999; Jones *et al.*, 2008). Importantly, landraces show phylogeographic patterns unencumbered by the overwriting effects of the intense plant breeding during the last century.

Extant landrace materials preserved in genebanks, however, also suffer from limitations to their usefulness in phylogeographic studies. In some areas, such as northern Europe, the number of available extant landraces is very low (Jones *et al.*, 2008). The passport data of these accessions are often limited to a country of origin and it is not always possible to determine whether or not older accessions are actual landraces or early cultivars. Extant accessions are, furthermore, maintained *ex situ* and at small population sizes, which unavoidably leads to genetic drift, in addition to the risk of contamination during propagation (Steiner *et al.*, 1997; Börner *et al.*, 2000; Parzies *et al.*, 2000; Chebotar *et al.*, 2002; Hagenblad *et al.*, 2012). Consequently, extant material, at least in certain geographic areas, may inadequately represent the genetic diversity once present in landraces, thereby obscuring phylogeographic patterns.

Few studies have reported fine-scale geographic structure of landrace crops (but see Pandey *et al.*, 2006; Yahiaoui *et al.*, 2007; Rodriguez *et al.*, 2012). This may, in part, be due to the lack of suitable landrace material. Fennoscandia (Norway, Sweden, Finland

and Denmark), however, provides unique opportunities to explore fine-scale geographic structure in crops and how it relates to known agrarian history. In Sweden, Finland and Norway, seed collections, compiled by agronomists during the late nineteenth century, mainly to be used as display objects, provide an alternative material to extant landraces (Leino *et al.*, 2009;Leino, 2010). Sampling was carried out 'on farm', and sampling location is frequently detailed down to the specific farmstead. Although documentation on the sampling method is not available, it is reasonable to assume that samples are representative for the farms where they were collected as the seed volumes are large and show no signs of seed sorting. The historical seeds are original samples and have, in contrast to extant material, not been subject to any change in genetic diversity since the sampling.

Fennoscandia holds the northernmost expansion of barley in the world, and encompasses large variation in climate, soil and light regimes. Historically, the Baltic Sea and the Bay of Bothnia between Sweden and Finland have facilitated trade in the region, while the Scandes mountain range between Norway and Sweden has been a natural barrier for both trade and agriculture (Flygare, 2011). Historical documents describe both trade regulations limiting seed import and crop failures leading to the need for long-range seed trade. A study of fine-scale geographic structure would allow the long-term genetic effects of such historical events to be assessed. Studying Swedish barley accessions from the historical collections, genotyped for 14 microsatellite markers, Leino and Hagenblad (2010) detected a north-south separation of genetic diversity and suggested two separate colonization routes into the country. The number of markers used did not, however, allow any detailed genetic structuring to be detected and the sampling limited to Sweden prevented broader conclusions concerning Fennoscandian barley from being drawn.

Here we report a study of the genetic structure of landrace barley sampled with a high geographic resolution across all of Fennoscandia. We have used high-throughput single nucleotide polymorphism (SNP genotyping to screen multiple individual seeds from a large number of primarily historical specimens. This allows us a population genomic approach to explore the details of the species' genetic structure at the northernmost limit of its distribution range.

MATERIALS AND METHODS

Plant Material

A total of 40 six-row barley accessions were studied (Table 1). Of these the majority, 31 accessions, were taken from the nineteenth century historical seed collections; Tromsø University Museum in Norway (TR, three accessions), Mustiala Agricultural College in Finland (MU, six accessions) and the Swedish Museum of Cultural History in Sweden (NM, 21 accessions) (Leino *et al.*, 2009; Leino, 2010). The seeds, which are no longer viable, were collected at harvest on farm in 1869 (TR), 1890s (MU) and 1896 (NM), respectively, except for NM264 that was collected in 1882. Accessions were chosen for best possible coverage of Fennoscandia (Figure 1, Table 1), and where geographic coverage was lacking in the seed collections, the historical accessions were complemented with nine extant accessions provided by the Nordic Genetic Resource Center (NGB) and the N.I. Vavilov Research Institute of Plant Industry (VIR). On the basis of passport data, seed jar labels and visual inspection, only hulled six-row spring barley landraces were chosen. Plant improvement of six-row barley for Fennoscandia did not begin until the 1920s (Osvald, 1959) and the historical material can therefore be considered to be genuine landraces.

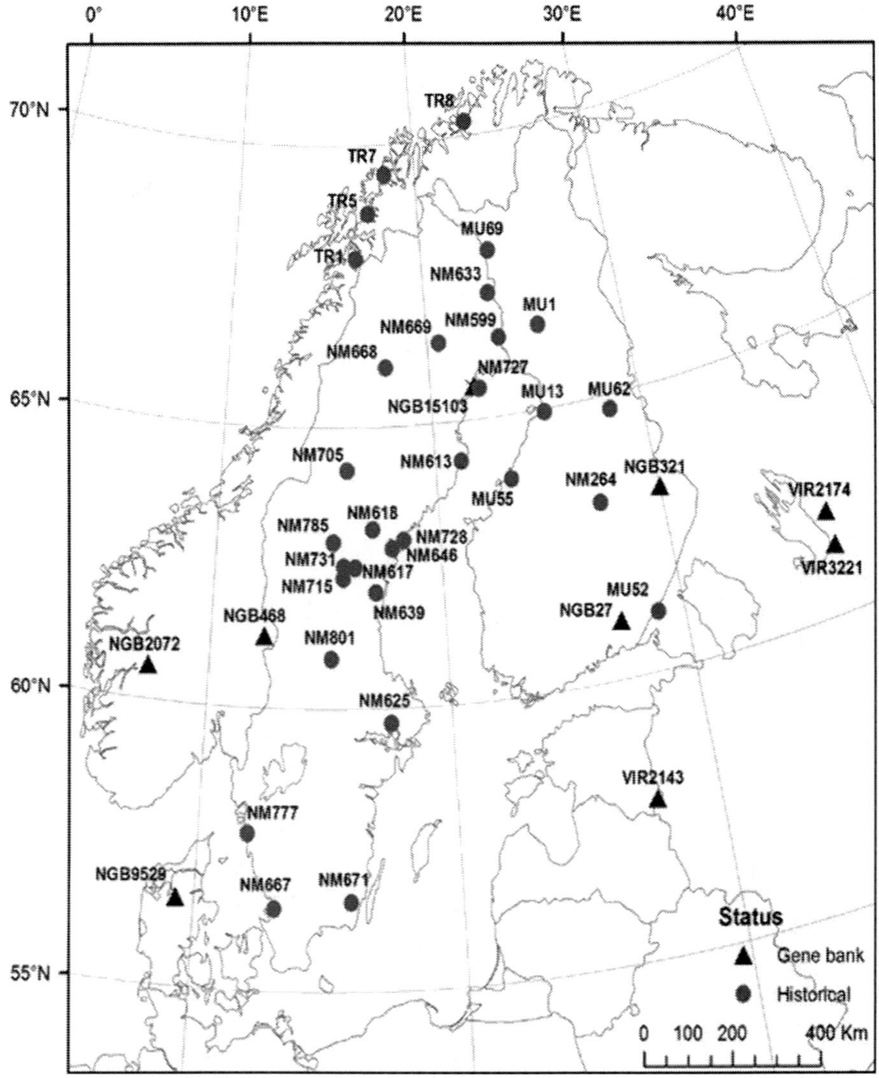

Figure 1: Geographic origin of the accessions and their conservation status. Country borders on the map are the borders of 2014.

Table 1: Geographic information and diversity data for the landrace accessions used in the study

Accession	Site of origin (Country)	Source	Lat. (N)	Long. (E)	NInd	NUGen	Nei's h	PC dispersion (Var)	Interchromosomal LD (r2)
MU1	Rovaniemi (FIN)	Historical	66.48	25.72	5	4	0.056	4.507 (2.544)	0.335
MU13	Oulunsalo (FIN)	Historical	64.93	25.40	6	5	0.044	3.766 (2.944)	0.358
MU52	Jääski (FIN/ RUS)	Historical	61.03	28.92	6	5	0.089	5.586 (2.917)	0.269
MU55	Vielvis/Kelviå (FIN)	Historical	63.85	23.45	6	6	0.083	5.542 (1.132)	0.255
MU62	Hyrynsalmi (FIN)	Historical	64.73	28.47	6	6	0.063	4.709 (2.247)	0.318
MU69	Muonionniska (FIN)	Historical	67.95	23.65	6	6	0.078	5.209 (0.471)	0.278
NGB15103	Luleå (SWE)	Extant	65.59	22.15	6	4	0.113	6.047 (6.537)	0.370
NGB2072	Finset (NOR)	Extant	60.60	7.50	6	3	0.004	0.925 (0.728)	0.760
NGB27	Sarkalahti (FIN)	Extant	61.03	27.33	6	5	0.085	5.508 (2.528)	0.278
NGB321	Törmälä (FIN)	Extant	63.18	30.02	6	6	0.180	7.786 (3.797)	0.366
NGB468	Trysil (NOR)	Extant	61.28	12.28	6	2	0.089	3.127 (20.952)	1.000
NGB9529	Lynderupgaard (DNK)	Extant	56.57	9.35	6	2	0.011	1.178 (2.917)	1.000

NM264	Mattila (FIN)	Historical	63.15	27.28	6	6	0.106	5.736 (0.437)	0.420
NM599	Matarengi (SWE)	Historical	66.38	23.65	6	5	0.068	4.866 (1.826)	0.340
NM613	Sundby (SWE)	Historical	64.28	21.22	6	6	0.086	5.577 (0.426)	0.248
NM617	Österkomsta (SWE)	Historical	62.50	16.17	6	6	0.086	5.708 (0.232)	0.205
NM618	Forssen (SWE)	Historical	63.17	16.99	6	6	0.106	6.339 (0.554)	0.230
NM625	Toppmyra (SWE)	Historical	59.73	17.55	6	6	0.067	4.964 (0.780)	0.263
NM633	Pajala (SWE)	Historical	67.20	23.37	5	2	0.013	1.455 (3.445)	1.000
NM639	Bergsjö (SWE)	Historical	62.05	17.06	5	5	0.070	5.273 (0.173)	0.255
NM646	Ramvik (SWE)	Historical	62.82	17.85	6	6	0.089	5.780 (0.780)	0.251
NM667	Nya Skottorp (SWE)	Historical	56.45	13.00	5	5	0.078	5.546 (0.535)	0.303
NM668	Kurrokveik (SWE)	Historical	66.05	17.88	6	6	0.069	5.122 (0.287)	0.213
NM669	Vuollerim (SWE)	Historical	66.42	20.62	6	6	0.065	4.927 (0.223)	0.209
NM671	Hylkebo (SWE)	Historical	56.58	15.85	6	6	0.074	5.300 (0.128)	0.200
NM705	Brattbäcken (SWE)	Historical	64.23	15.87	6	5	0.073	5.121 (1.753)	0.288
NM715	Ransjö (SWE)	Historical	62.31	15.65	6	6	0.079	5.448 (0.198)	0.202

NM727	Sandön (SWE)	Historical	65.53	22.40	6	6	0.064	4.913 (0.217)	0.213
NM728	Omne (SWE)	Historical	62.96	18.37	6	6	0.071	5.182 (0.217)	0.209
NM731	Ånge (SWE)	Historical	62.52	15.66	5	5	0.081	5.638 (0.304)	0.272
NM777	Assmundstorp (SWE)	Historical	57.77	11.92	6	6	0.081	5.496 (0.990)	0.237
NM785	Landsom (SWE)	Historical	62.95	15.22	6	6	0.059	4.692 (0.278)	0.215
NM801	Stumnäs (SWE)	Historical	60.88	15.12	6	6	0.082	5.507 (0.799)	0.256
TR1	Storfjord (NOR)	Historical	68.00	16.50	5	5	0.049	4.301 (0.641)	0.330
TR5	Ibestad (NOR)	Historical	68.80	17.25	5	5	0.044	4.193 (0.336)	0.287
TR7	Balsfjord (NOR)	Historical	69.47	18.25	5	5	0.056	4.683 (0.792)	0.298
TR8	Komagfjord (NOR)	Historical	70.25	23.22	5	5	0.074	5.299 (0.285)	0.301
Vav2143	Estonia (EST)	Extant	57.82	27.60	6	4	0.198	7.160 (17.144)	0.492
Vav2174	Karelia (RUS)	Extant	61.82	36.67	6	4	0.275	8.985 (19.290)	0.543
Vav3221	Karelia (RUS)	Extant	61.22	36.63	6	3	0.188	5.921 (29.080)	0.699
						Average	0.083	6.816 (2.600)	0.364

Accessions abbreviated with MU are from the seed collection at the Mustiala Agricultural College, NGB from the Nordic Genetic Resource Center, NM from the seed collection at the Swedish Museum of Cultural History, TR from the seed collection at the Tromsø University Museum and VIR are from the N.I. Vavilov Research Institute of Plant Industry. Countries are abbreviated according to ISO 3166-1 alpha-3. Source indicates whether the accession was collected from a genebank (extant) or from historical seed archives (historical). Latitude and longitude are given in decimal degrees. N_{Ind} indicates the number of individual seeds genotyped and N_{UGen} indicates the number of unique genotypes within the accession. PC dispersion is the average distance in PC-space between each pair of individual seeds in the accession and is given along with its variance (in parenthesis). All principal components were included when calculating the PC dispersion.

DNA Analysis

DNA was extracted from six individual seeds from each accession using FastDNA Spin Kits and the FastPrep Instrument (MP Biochemicals, Solon, OH, USA). Extractions were performed at a laboratory separate from that where SNP genotyping was carried out to reduce the risk of contamination. A negative control was included in each extraction series. Two present day cultivars (cv. 'Morex' and cv. 'Rolfi') were used as positive controls and three negative extraction controls were included in the SNP genotyping. SNP data were generated using an Illumina Golden Gate assay (Illumina Inc., San Diego, CA, USA), with the C-384 SNP set designed for optimal diversity for European barley cultivars, as developed by Moragues *et al.* (2010). The resulting data were processed and studied with the Bead Studio 3.1.3.0 software packager (Illumina Inc., San Diego, CA, USA). To verify that repeatable and authentic SNP calling could be performed on historical material by the assay, four DNA extracts from kernels of the same 100-year-old ear (NM76) were genotyped. Additionally, DNA extracts from historical samples of the cultivars 'Gull' (NM52) and 'Princess' (NM60) and extant material of the same cultivars (NGB1480 and NGB9424) were compared. To evaluate ascertainment bias, folded minor allele frequency spectra were generated for the full data and for three regional subsets of the data. Accessions with an origin north of the 65th parallel were categorized as 'North', accessions with an origin between the 60th and 65th parallel as 'Mid' and accessions with an origin south of the 60th parallel categorized as 'South'.

Linkage Disequilibrium

Linkage disequilibrium (LD) was calculated as r^2 (Hill and Robertson, 1968) using a purpose-written Perl script. Intrachromosomal LD was calculated for pairs of polymorphic loci residing on the same chromosome and interchromosomal LD was calculated for pairs of polymorphic loci located on different chromosomes. LD was calculated both across all individuals and for the individuals of each accession.

Statistical Analysis

Principal component analysis (PCA) was performed using the *prcomp* function in the statistical software *R* (R Development Core Team, 2013, version 3.0.2) to visualize both within accession diversity and structure between accessions. The SNP data were analyzed both as individuals and on an accession level. For the individual level, each homozygous SNP was treated as either 1 or 0 and missing data were replaced with the allele frequency in the full dataset of the allele designated as '1'. For the accession level PCA, allele frequencies of each accession for each of the SNPs were calculated and treated as independent variables. A measure of genetic relatedness between individuals within accessions, based on principal components, was calculated using R. This measure, called PC dispersion, was the mean pairwise distance in PC-space between individuals within accessions. Data of all principal components for each individual in an accession were used as coordinates in a multidimensional space and the average distance between individuals belonging to the same accession in this multidimensional space was calculated.

The software structure (Pritchard *et al.*, 2000; Falush *et al.*, 2003, version 2.3.3) was used to cluster accessions into populations. As suggested by Nordborg *et al.*(2005) for selfing species, and applied in other studies (Pandey *et al.*, 2006; Leino and Hagenblad, 2010; Leino *et al.*, 2013), we analyzed data as haploid, treating heterozygous loci as missing data. *Structure* simulations were carried out using an admixture model. Burn-in period was set to 25 000 iterations and estimations were based on 50 000 iterations. The simulations were repeated 20 times for K-values of 1–10. The choice of relevant numbers of clusters was guided by calculating ΔK using the method presented in Evanno *et al.* (2005) and the change in H′ from CLUMPP. To properly evaluate multimodality of the *structure* output, the 20 repeats for each K from the *structure* simulations were merged using the CLUMPP software (Jakobsson and Rosenberg, 2007). CLUMPP was used with the Greedy Algorithm method and the results were visualized using the Distruct v1.1 software (Rosenberg, 2004). The same procedure and settings were used for all analyses of genetic structure.

To verify that our results were not influenced by the *structure* assumption of Hardy-Weinberg equilibrium, discriminant analysis of principal components (DAPC) was used as an alternative method of evaluating population clustering. DAPC is a multivariate method included in the *Adegenet* R package (Jombart *et al.*, 2010) and requires no prior assumption of the underlying population genetic model. The method was applied both to the full data and to a subset consisting of only the historical accessions. All principal components were utilized for prior group clustering and DAPC analysis used a subset of 50 PCs to prevent over-fitting.

Within-accession genetic diversity was assessed by calculating Nei's h ($h=1-\Sigma\ p_i^2$, where p_i is the frequency of the ith allele) for each accession (Nei, 1973). This was carried out both for individual SNPs and for haplotypes (length 2–10 SNPs) consisting of merged neighboring SNPs. The distribution of genetic diversity was further explored by calculating pairwise F_{ST} values (Weir and Cockerham, 1984) between the accessions. Pairwise F_{ST} values were also analyzed between pairs of accessions in the three previously defined latitudinal groups. Average genetic diversity (Nei's h) for the genotyped SNPs and pairwise F_{ST} between accessions and groups of accessions were calculated using the *Arlequin 3.5* software (Excoffier and Lischer, 2010). The significances of F_{ST} values were estimated with permutation tests (1000 permutations). The relative effect of gene flow and drift was also analyzed by plotting pairwise F_{ST} values against geographic distances (Hutchison and Templeton, 1999). Geographic distances between accessions were calculated in *R* using the haversine formula. Total genotype sharing, that is when two individuals had identical genotypes at all loci, excluding missing data, was also determined using *Arlequin 3.5*.

Geographic Visualization

Maps for geographic visualization of genetic structure were created using ArcGIS (ESRI, 2011), with geographic data available through the 'ESRI data and maps v. 9.3' database (2008).

RESULTS

SNP Calling in Historical and Extant Barley Accessions

We have measured genetic diversity in more than 115-year-old historical and extant barley using an Illumina Golden Gate SNP assay. DNA from barley seeds in the historical collections is degraded to fragments of typically 100–200 bp and yield 200 ng mg^{-1} seed, compared with extant seeds where DNA yield is typically four times higher and DNA length is above 10 000 bp (Leino *et al.*, 2009). DNA concentrations in the extracts used for SNP genotyping were in the range of 20–65 ng µl^{-1}. Although DNA concentration is low, it proved to be of sufficient quality for successful genotyping and whole genome amplification using Illustra GenomiPhi (GE Healthcare Life Sciences, Buckinghamshire, UK), of eight test individuals did not improve genotyping success.

To verify that repeatable and accurate signals could be obtained from the historical material, four replicate DNAs extracted from seeds from the same 100-year-old ear were analyzed. As barley is highly inbreeding, the seeds can be expected to have very high genetic similarity and SNP calling from the four extracts were indeed identical, demonstrating repeatability. To further verify that extant and historical material could produce comparable genotype scores, historical and extant accessions of two cultivars, 'Gull' and 'Princess' were tested. Even though it is possible that the cultivars have changed slightly over the past 100 years, we nonetheless expect, if our genotyping method yields accurate results, historical and extant samples from the same cultivars to be more similar than when comparing the historical and the extant accessions, respectively, with each other. Comparing historical and extant accessions of the same cultivar resulted on average in 91.8% identical scores (100.0% and 85.4% for 'Gull' and 'Princess' respectively), whereas comparing extant material of different cultivars resulted in 63.2% identical scores and comparing historical material resulted in 71.7% identical scores. We thus concluded that SNP calling of the historical material was sufficiently accurate and not influenced by the nature of aged DNA.

A total of 231 individuals from 40 different accessions originating across Fennoscandia were obtained from historical seed collections (31 accessions) and genebanks (9 accessions) (Table 1, Figure 1). These, together with three negative extraction controls were assayed for 384 SNPs. Neither of the extraction controls yielded detectable SNP signals verifying the absence of contamination. Loci scored as heterozygous within an individual were extremely rare in the dataset, 0.076% averaged across all loci and individuals. Because of difficulties of separating unclear marker distinction from actual heterozygotes, such loci were scored as missing data. Of the 384 SNPs assayed, 63 were excluded because of high levels (>15%) of missing data and one because of being monomorphic, leaving 320 SNPs to be used for analyses. Additionally, nine individuals with more than 40% missing data were excluded from further analysis leaving an average 1.6% missing SNP data per individual (on average, 2.01% for the historical material and 0.09% for the extant material). Success rates of individual SNP markers are reported as supporting information

Minor allele frequency spectra for the total dataset and the predefined latitudinal groups showed that the spectra for the subgroups differ somewhat from each other and that of the total dataset. This is likely due to stochastic effects as the underlying data for the subgroups are based on fewer individuals. In all spectra, the frequency of the minor allele was in most cases 5%or less. The number of SNPs within each category of minor allele frequency was becoming increasingly lower as the minor allele frequency approached 50%, as expected under the basic model (see Nielsen *et al.*, 2004).

LD in Fennoscandian Barley does not vary with Latitude

LD between pairs of polymorphic loci was calculated as r^2. Average interchromosomal LD was for the complete dataset 0.0664 but with a skewed distribution (median 0.012). Intrachromosomal LD declined with the distance between markers, but r^2 values of 0.4 and more could still be observed between markers 100 cM of more apart. We also calculated interchromosomal LD for each accession separately (Table 1). Average interchromosomal LD ranged from 0.2 for NM671 to 1 for NGB468, NM633 and NGB9529 (average 0.364) but was not significantly correlated with latitude of the accession (P=0.247). We also compared LD values between accessions from each of three predetermined latitudinal groups: South, Mid and North. None of the three geographical groups South, Mid and North differed significantly in their LD from each other (two-sided t-test, all P>0.5). LD was, however, higher in extant material (average 0.612) than in historical (average 0.293) accessions (two-sided$t$-test, P<0.001).

Both Genetic Diversity and Total Genotype Sharing is Higher within Extant Landraces Compared with Historical Accessions

To assess the genetic composition of the accessions, both Nei's h and the amount of total genotype sharing within accessions were calculated. Within-accession genetic diversity ranged from 0.005 in the extant Norwegian landrace NGB2072 to 0.331 in the extant Russian landrace VIR2174 with an average within-accession diversity of 0.101 (Table 1, Figure 2). A genetic bottleneck leading to loss of genetic diversity could be expected in populations migrating northwards. However, within-accession genetic diversity was not found to be significantly correlated with latitude (r^2=0.11672, P=0.205). The average within-accession genetic diversity of the historical accessions (0.087) was significantly lower than extant accessions (0.152) (two-tailed unpaired t-test, P<0.001).

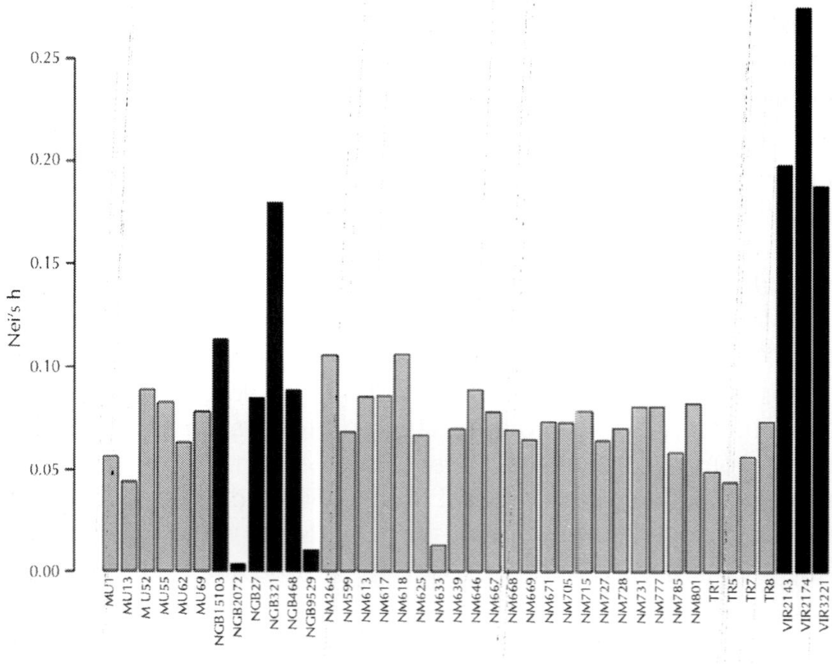

Figure 2: Within-accession genetic diversity measured as average genetic diversity (Nei's h) for the genotyped SNPs of each accession. Extant accessions are displayed as black bars and historical accessions are displayed as grey bars.

Ascertainment bias can affect comparisons of genetic diversity between unascertained and ascertained populations, or populations similar to ascertained populations. The effect of ascertainment bias can be alleviated by combining SNPs into haplotypes (Conrad *et al.*, 2006; Oliveira *et al.*, 2014). For this reason, Nei's h was also calculated for haplotypes of length 2–10. The haplotype diversities suggested little effect of ascertainment bias. The accessions with the highest and lowest diversity remained the same for all lengths of haplotypes and most differences in relative rank were minor. The few accessions that showed a marked change in relative diversity could be explained by an increase of missing data as the haplotypes were merged.

All but one extant accession (88.9%) included individuals sharing identical total genotypes (that is, two or more individuals within the same accession having all genotyped SNPs scored as identical), something that was only found in 6 out of 31 accessions (19.4%) among the historical material (Table 1). It is likely that the amount of total genotype sharing in the historical accessions is even lower, as missing data are more common, increasing the likelihood of individuals being classified as identical. It is worth noting that the three accessions with the highest genetic diversity, VIR2143, VIR2174 and VIR3221, also have a high degree of total genotype sharing, meaning they consist of a few, but very different, lines (Table 1). Although most of the historical accessions are more variable than the extant material, it should be noted that some of the individuals share total genotypes not only within accessions but also between accessions. This occurs in the accessions MU13, MU69, NM633 and NM668, all of which originate from relatively nearby places in northern Finland and northern Sweden.

Genetic Diversity in Extant Landraces is inflated by a Few Individuals with Distinctly Different Genotypes

We explored the distribution of genetic diversity by PCA. In the accession level PCA, the Estonian accession, VIR2143, clustered separately from all other accessions along PC2, whereas the Karelian accessions appeared closer to the Fennoscandian material (Figure 3a). There was, however, little indication of accessions clustering according to country of origin (Figure 3a). Classifying accessions according to latitudinal groups described above revealed a certain amount of clustering according to latitude along the first PC (Figure 3b), a pattern that was made clearer when omitting extant accessions (Figure 3c).

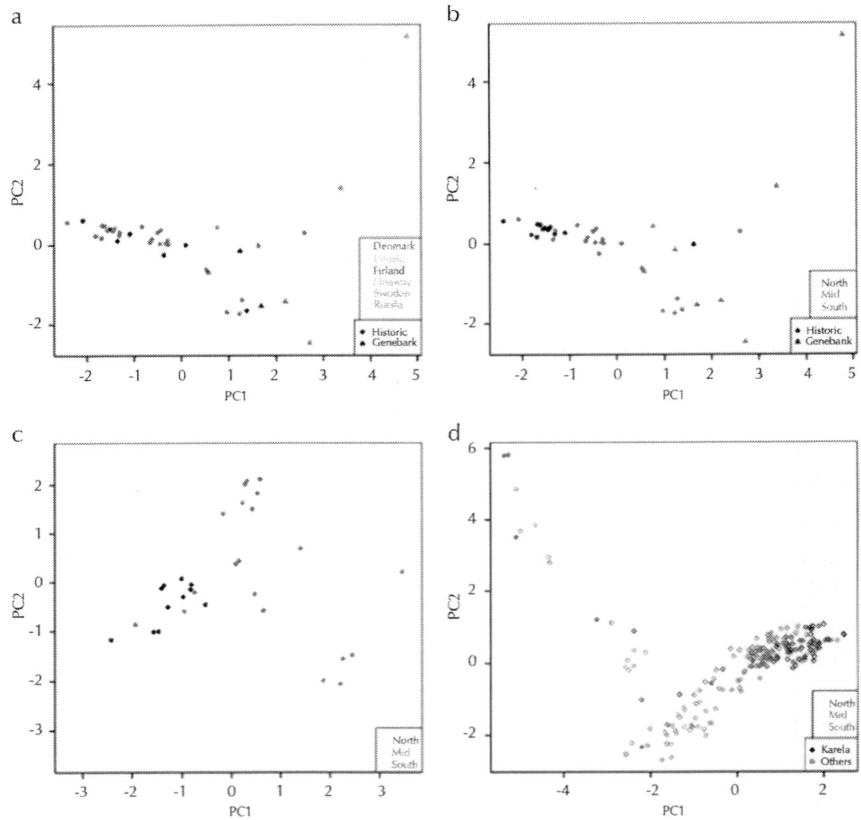

Figure 3: PCA of accessions and individuals. (a) PCA of 40 accessions of landrace barley on accession level. Accessions are divided by color depending on country of origin according to present day national borders and by symbol according to seed source. PC1 and PC2 explain 22.88% and 11.43% of the total variation, respectively. (b) PCA of the same data as in (a) presented according to latitude. Accessions with an origin north of 65° N ('North') are shown in blue, accessions from between 60° N and 65° N ('Mid') in red and accessions south of 60° N ('South') in black. Symbols are as in (a). (c) PCA of historical accessions only. PC1 and PC2 explain 25.00% and 17.57% of the total variation, respectively. Color coding is as in (b). (d) PCA of 321 individuals from 40 accessions of landrace barley, performed on an individual level. PC1 and PC2 explain 12.25% and 7.46% of the total variation, respectively. Color coding is as in (b). Individuals from the two Karelian accessions are marked with filled circles to visualize their within-accession differentiation.

The geographic distances between accessions were well explained by the distances in PC space (r^2=0.265, P<0.001). Analyzing latitude and longitude separately as explanatory variables showed that latitude explained the distances between accessions in PC space much better (r^2=0.224, P<0.001) than longitude (r^2=0.0366, P<0.001). This is noteworthy as the geographical range of the study area is approximately equal in both directions.

The individual level PCA showed that specimens of the Karelian accessions (VIR2174 and VIR3221) were widely distributed along both PC1 and PC2 with individuals being either very similar to the Fennoscandian material or highly divergent (Figure 3d). To explore this further, a proxy of the genetic variation within accessions, PC dispersion, was quantified from the PCA results using mean pairwise distances in PC-space (Table 1). These values ranged from 1.063 for the Norwegian accession NGB2072 to 9.319 for the Karelian accession VIR2174 and were highly correlated with the within-accession genetic diversity (r=0.867,P<0.001). The variance of the PC dispersion statistic provided additional information to the mean PC dispersion as it is strongly inflated where the genetic variation between individuals within an accession is uneven, such as when a sample contains seed from mixed sources. Thus, a high PC dispersion variance indicates accessions that may have been subject to seed mixing during rejuvenation. For example, the accession NGB468 shows a high variance in PC dispersion (24.048) because of a single individual which differs greatly, whereas the remainder of the material contains multiple identical total genotypes (N_{UGen}/N_{Ind}=0.33); this could not be detected from the genetic diversity index (h=0.101), which is in fact below average. High PC dispersion variance was found in several extant accessions but none of the historical accessions (Table 1).

Differences in the genetic diversity of extant and historical landraces were also evaluated by comparing extant and historical accessions from the same geographic origin to rule out any geographical effects on diversity. The two pairs of extant and historical accessions with the shortest geographic distance are NGB15103 (extant)—NM727 (historical) and NGB27 (extant)—MU52 (historical) where the accessions NGB15103 and NM727 originate from nearly the same site. The extant NGB15103 had significantly higher genetic diversity (0.136 vs 0.077, two-tailed unpaired *t*-test, P<0.001), fewer unique total genotypes (four vs six) and a higher PC dispersion (6.680 vs 0.217) compared with the historical NM727, which could indicate that it is a mixture of seeds with different origins rather than a genuine landrace. In contrast, the genetic diversity of the accessions NGB27 and MU52 were not significantly different (Nei's h=0.102 vs 0.107 for NGB27 and MU52, respectively, two-tailed unpaired *t*-test, *P*=0.76) and each accession had two individuals with total genotype sharing among six individuals. The variances of the PC dispersion of NGB27 and MU52 were also quite similar, being 2.528 in NGB27 and 1.611 in MU52. F_{ST} values further indicated differences between NGB15103 and NM727 with a high and significant F_{ST} (F_{ST}=0.422, *P*<0.001). NGB27 and MU52 in contrast had a low and non-significant F_{ST} value (F_{ST}=0.077, *P*=0.144).

A Latitudinal Structure of Genetic Diversity

We explored geographic structuring of the accessions with the software*structure*. We initially analyzed the complete dataset where Evanno's ΔK suggested that a two-cluster model best described the data and the H′ values from CLUMPP indicated either two or three populations with equal support. Although ΔK cannot be calculated for k=1, log-likelihood values calculated by *structure* were consistently much higher for k>1 than k=1, which, together with the distinct and reproducible clustering patterns, indicates that population structure is best described by more than one cluster. The two-cluster model separated the entire VIR2143 accession and individuals from the accessions VIR2174, VIR3221, NGB321 and NGB468 (Figure 4a). Increasing the number of clusters to three (data not shown) divided the Fennoscandian accessions in a north and south cluster.

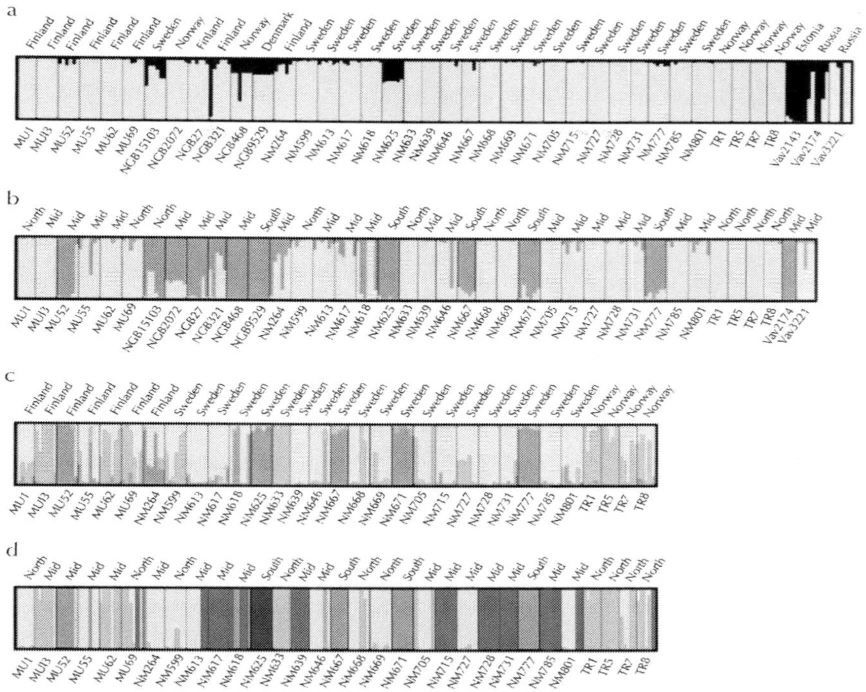

Figure 4: (a–c) Genetic structure from 20 individual *structure* simulations merged with the CLUMPP software. (a) K=2, complete dataset. (b) K=2, subset with the minor (black) cluster from (b) removed. (c) K=3, historical accessions only. (d) Output from DAPC analysis, visualized with the Distruct software, K=5, historical accessions only. A full color version of this figure is available at the *Heredity*journal online.

To properly evaluate any substructure that may have been obscured by large difference between the main clusters, and based on the information from the PCA and the *structure* analysis, we created two subsets of accessions. The first subset excluded the divergent cluster from the first *structure* analysis by removing individuals that clustered to the minor cluster in the k=2 model to an extent of 90% or more (shown as black in Figure 4a). For this subset, ΔK again supported k=2, while H′ indicated k=2, but with nearly as high support for k=4. The two-cluster model of the first subset divides Fennoscandia into a northern and southern cluster, across national boundaries (Figure 4b). The four-cluster model adds two additional, latitudinal, clusters (data not shown).

The second subset consisted of only historical accessions to assess whether the inclusion of the extant material in general has an effect on the genetic structure. Indeed, the clusters in Fennoscandia proved to be much more distinct with the omission of extant accessions (Figure 4c). For the historical seed subset, the ΔK indicated k=2 while H′ was highest for k=3 and nearly as high for k=2 and k=6 (. The three-cluster model divided the accessions into three latitudinal groups, the southernmost encompassing the southern third of Sweden and the southern tip of Finland, a middle group of accessions stretching eastward across central Sweden and Finland and a northernmost group with accessions from northern Norway, northern Finland and Sweden along the border to Finland (Figures 4c and 5).

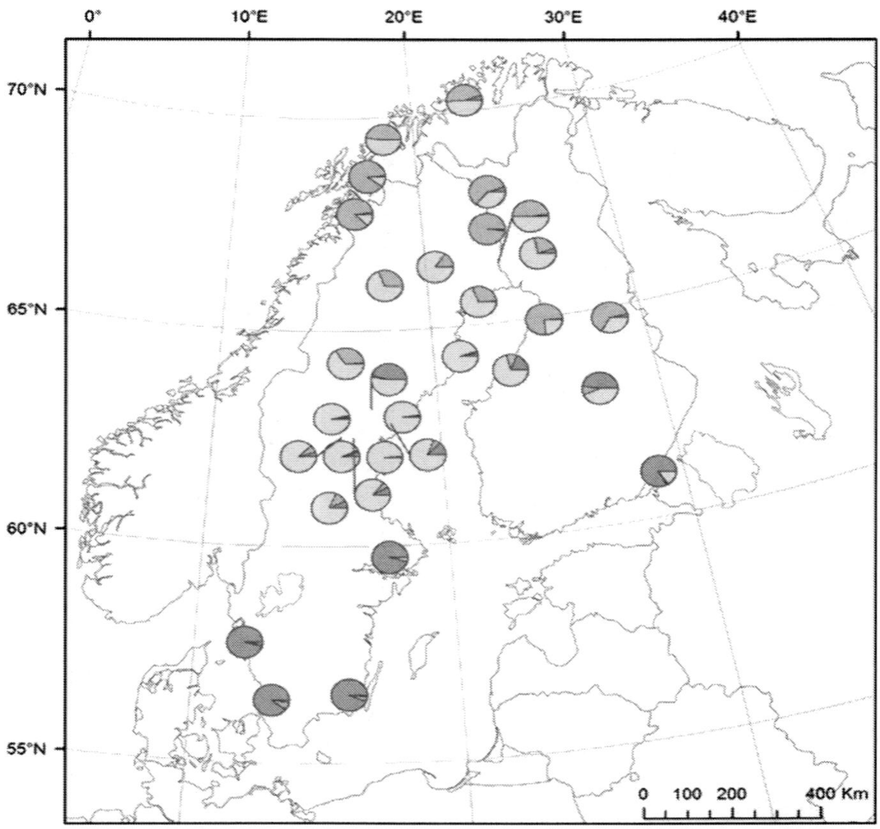

Figure 5: Geographic visualization of *structure* results from 20 simulations with K=3, joined together with CLUMPP and visualized with ArcMap 10.

Clusters are colored as in Figure 4. A full color version of this figure is available at the *Heredity* journal online.

We also evaluated the data with discriminant analysis of principal components (DAPC), an alternative method not assuming Hardy-Weinberg equilibrium. These DAPC analyses supported k≈9–10 for the complete dataset and clustering for k=2 to k=10 were nearly identical to those of the*structure* analyses; for all levels of k, the structuring observed was primarily latitudinally distributed. For the historical subset, DAPC clearly supported k=5 (Figure 4d) and the overall structure was again very similar to that of the k=3 model from *structure*, with clear latitudinal structuring, albeit separating the NM625 into its own cluster and most Swedish accessions between the 60th and 65th parallel into yet another cluster (Figure 4d).

F_{ST} values were calculated between all pairs of accessions and 89.6% of the pairwise F_{ST} values were significant at the level $P<0.05$. Pairwise F_{ST} values and geographic distances were moderately but significantly correlated (r=0.497, $P<0.001$). A scatter plot visualized this correlation and also indicates increasingly varying F_{ST} values as distances increase (Figure 6). The resulting correlation suggests a lack of regional equilibrium between gene flow and drift (see Hutchison and Templeton, 1999). For the landrace barley, it would appear that gene flow is more effective at shorter distances and drift is more influential at greater distances of geographic separation.

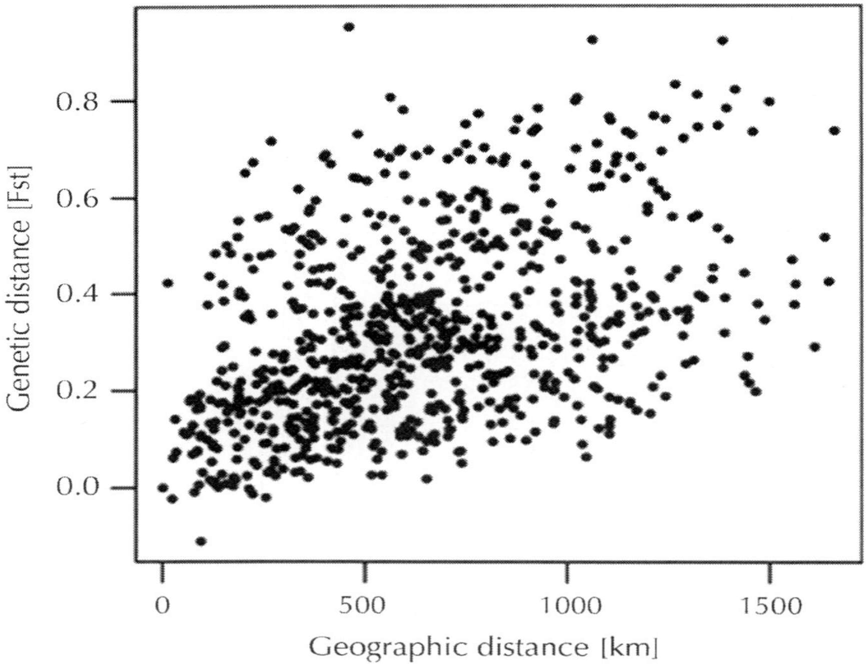

Figure 6: All pairwise F_{ST} values plotted against geographic distance between the accessions.

After excluding the extant material, we averaged pairwise F_{ST} for comparisons within and between the previously defined latitudinal regions 'North', 'mid' and 'South' (Table 2). In the three groups, the highest average pairwise F_{ST} was found between accessions in the 'South' group, which could be an indication of a higher genetic diversity within this latitudinal group. The corresponding statistic was lower for the 'North' and 'Mid' group where average F_{ST} values were approximately the same. The pairwise comparisons between groups revealed that the differences between accessions in the 'North' and 'Mid' group were less than the difference of either group from the 'South' group. The F_{ST} averages for the different comparisons between regions were all significantly different from each other (two-tailed unpaired *t*-test, $P<0.001$).

Table 2: Averages of pairwise F_{ST} comparisons within accessions (diagonal) and between latitudinal groups of accessions

	North	Mid	South
North	0.140 (0.002)		
Mid	0.238 (0.009)	0.199 (0.001)	
South	0.455 (0.017)	0.387 (0.016)	0.338 (0.040)

Standard errors of the averages are reported in brackets. All averages differed significantly from each other.

DISCUSSION

Resolving the historical spread of agricultural crops is a challenging task requiring combined archaeological, botanical and historical evidence. Recently, phylogeographic studies have added substantially to the understanding of crop history (Van Heerwaarden *et al.*, 2011; Oliveira *et al.*, 2012; Roullier *et al.*, 2013). However, if a study is performed on a finer geographical scale, more power in terms of genetic markers and number of individuals is needed. Additionally, the use of landraces with strong genetic integrity becomes even more important. When these requirements are fulfilled, as shown in the present study, patterns of historical crop spread can become visible.

Phylogeographic studies of landrace crops have often been restricted to single individuals (for example, Saisho and Purugganan, 2007; Isaac *et al.*, 2010; Van Heerwaarden *et al.*, 2011) or aggregate samples from accessions, using pooling schemes (for example, Hunt *et al.*, 2011; Jones *et al.*, 2011; Oliveira *et al.*, 2012). This allows a much higher number of accessions to be studied, but comes at the cost of ignoring within-accession genetic diversity. This is unfortunate, both because knowledge on within-accession diversity is valuable *per se*, but even more so as within-population diversity aids in the proper identification of population structure (Fogelqvist *et al.*, 2010; Lascoux and Petit, 2010).

The trade-off between the number of individuals per accession studied and number of accessions that can be included in a study means a careful balance must be struck. Although a higher number of individuals aids in the identification of population structure, the overall scatter obtained here when plotting F_{ST} vs geographic distance (Figure 6) illustrates the need also for a large number of accessions to determine genetic structure of the species. This is particularly the case when the distance between sampling locations is high. The increasing variance in F_{ST} values as distance increase suggests that a dense sampling of populations will facilitate the evaluation of genetic structure. Ideally, a minimum number of individuals, such as the 5–10 individuals suggested by Fogelqvist *et al.*(2010), should be sampled while ensuring that the necessary number of populations can still be studied.

Our analysis of the between-individual distribution of within-accession diversity suggests that NGB468, VIR3221 and VIR2174 may be of mixed origin. Although the genetic diversity of these accessions is extremely high, individuals sharing the same total genotype are common, that is, there are several distinct total genotypes each shared by more than one individual, as would be expected in seed mixtures or lineages descending from seed mixtures. Additionally, although some individuals share origin with the Nordic group, others appear to have a different origin (Figure 3d, Karelia). It is not possible to determine whether this mixture is a result of contamination during genebank maintenance or if the mixture was present at the site at the time of collection. If the original collection occurred after a recent crop failure event, a part of the seed could have been recently imported from a different area. Further testing of material from areas suggested as potential trade partners in historical records will be needed to elucidate this.

Although the benefit of genebank conservation for plant breeders is unquestionable (reviewed by De Carvalho *et al.*, 2013), concerns have been raised regarding the use of extant landrace material for studying questions regarding crop evolution (Lister *et al.*, 2009; Hagenblad *et al.*, 2012, Leino *et al.*, 2013;Roullier *et al.*, 2013). Genetic drift, selection and contamination are all processes that can lead to changes in the genetic identity of landraces during *ex situ*regenerations. Recently collected *in situ* preserved landraces, when available, can instead have been affected by twentieth century large-scale seed trade and movement. Our results suggest that some of these processes may well have had an effect on the genetic composition of our extant material. The extant material contains accessions that are both markedly less diverse (such as NGB2072 and NGB9529), possibly a consequence of strong genetic drift, and clearly more variable (for example, NGB321, VIR3221, VIR2143 and VIR2174), which could be due to contamination, than all or most of the historical accessions (Figure 2). However, it must be noted that the extant material in this study was chosen to cover geographic areas from which historical material was not available. This means that both age and geography may play a part in any comparisons between extant and historical material. From the historical samples, we draw the conclusion that the within-accession genetic diversity of Fennoscandian landrace barley before the introduction of modern plant improvement was, for these SNP markers, typically somewhere between 0.05 and 0.1.

The SNP markers used were originally ascertained on smaller discovery panels of barley cultivars (Rostoks *et al.*, 2005, 2006; Close *et al.*, 2009), and hence ascertainment bias may influence the results. Moragues *et al.* (2010) tested a set of 1536 SNP markers on a large set of 500 cultivars and 169 landraces, evaluating ascertainment bias and its effect on diversity and proposed two reduced sets of 384 SNPs optimized for European cultivars and Syrian and Jordanian landraces, respectively. Of these, we have used the one optimized for European cultivars. Our material is restricted to landraces and although ascertainment bias could render inferences of genetic variation outside the area of the study somewhat inaccurate, it is unlikely to have an effect on the genetic structure or comparisons of population parameters within the area. We also note that neither the minor allele frequency spectra nor the genetic diversity of the haplotype groups indicate ascertainment bias as a major issue in this dataset.

Genetic analyses of historical specimens are a desirable alternative to steer clear of some of the issues associated with extant landraces (Jones *et al.*, 2008; Lister*et al.*, 2010, Hagenblad *et al.*, 2012). However, the difficulties involved in obtaining sufficient numbers of individuals and quality of DNA for such studies means the number of investigations reporting within-accession diversity in historical material is even lower than that of extant landraces (but see Leino and Hagenblad, 2010; Hagenblad *et al.*, 2012; Leino *et al.*, 2013). Additionally, the degraded quality of DNA in most historical samples means that genetic analysis can be difficult and in many cases a relatively low number of genetic markers, usually chloroplast or mitochondrial and sometimes microsatellite markers have been analyzed (reviewed by Palmer *et al.*, 2012).

Here we show that seeds more than 100 years old, when stored under favorable conditions, contain sufficient DNA to be analyzed with high-throughput SNP genotyping. SNP genotyping of historical specimens has previously been reported from animal samples, such as cattle (for example, Svensson *et al.*, 2007) and salmon (for example, Johnston *et al.*, 2013), and in small-scale studies of plants (for example, Lister *et al.*, 2013). To our knowledge, this is the first report where historical plant samples have been analyzed by large-scale SNP genotyping, with more than one individual per accession and a high number of markers. Should this prove to be possible also for other material, it will expand the field of population genetics of historical plant specimens into that of population genomics. Historical seed collections such as those at the Swedish Museum of Cultural History, the Mustiala Agricultural College and Tromsø University Museum, are likely to prove unusually suited for these types of studies as the number of seeds in collections allows genomic analysis on a population scale.

Our genomic scale analysis allowed us to investigate issues such as LD. We found limited evidence of breakdown of intrachromosomal LD, contrasting with previous findings in both wild barley (Morrell *et al.*, 2005; Caldwell*et al.*, 2006) and barley cultivars (Caldwell *et al.*, 2006) but showing striking similarities to the landrace dataset from Syria and Jordan studied by Caldwell *et al.* (2006). Average intrachromosomal LD was also similar to that reported for landrace barley from Syria and Jordan by Russell *et al.* (2011) but higher than in wild barley from the same area and lower than in a worldwide set of both two- and six-row barley cultivars (Malysheva-Otto *et al.*, 2006). Our results thus suggest that a similar genetic structure is maintained between landraces from the area of domestication and those from Fennoscandia. Additionally, landraces in general seem to show more intrachromosomal LD than wild barley, but less than improved cultivars. Our interchromosomal LD was, however, an order of magnitude higher than the one found in populations of Sardinian landrace barley (Rodriguez*et al.*, 2012). Future studies will reveal whether this is an effect of the sampling size or the result of differences in population structure and gene flow between different areas. The similar levels of within-population LD in the different geographical groups suggest that levels of outcrossing and gene flow between accessions are of similar magnitude across Fennoscandia.

The spatial genetic structure of extant barley has been described earlier both on a worldwide scale (Malysheva-Otto *et al.*, 2006; Saisho and Purugganan, 2007) and across Europe (Jones *et al.*, 2011, 2012) using microsatellite markers. With only 14 microsatellite markers genotyped in landrace barley from historical seed collections, Leino and Hagenblad (2010) were able to show geographic structuring in barley on a much finer scale—within Sweden. Six-row barley landraces from the far north were genetically distinct from those of the rest of Sweden. This tentatively suggested two separate routes of migration into Sweden, but raised questions regarding the phylogeographic patterns of barley in the neighboring Fennoscandian countries.

Our analysis of 384 SNPs allowed us to verify the far northern Swedish population detected by Leino and Hagenblad (2010). We could further show that the cluster is distributed across the whole far-north Fennoscandia. Additionally, althoughLeino and Hagenblad (2010) could only detect a single far-north cluster, the increased power from the higher number of markers in this study presents a more detailed picture. In addition to the far-north cluster, two or possibly three more latitudinal clusters can be detected, all being shared across the longitudinal breadth of Fennoscandia. The far northern group detected by Leino and Hagenblad (2010) thus seemed not to be a Finnish group migrating into northern Sweden but the northernmost of a set of several latitudinally structured clusters shared across Fennoscandia. These results contrast with several historical records on seed movement within the area.

When analyzing the subset with only historical material the most conspicuous detail of the three-cluster *structure* model and the five-cluster is the far-northern group (depicted in turquoise/light grey in Figures 4 and 5). It stretches from Tromsø in the far north of Norway reaching east through the northernmost Sweden and Finland. This connection between northern Norway, northern Sweden and Finland is particularly fascinating as barley cultivation is not continuous in the area (Flygare, 2011), but interrupted by the Scandes mountain range. The genetic similarity instead appears to be a result of international seed trade over the mountain range. Such a trade in the far north of

Fennoscandia is noteworthy because mercantilist policies during the period 1593–1788 restricted foreign seed import into Norway, at that time under Danish rule (Herstad, 2000). This culminated with a state monopoly on seed trade in the period 1735–1788 for northern and western Norway (Lunden, 2004). It is, however, known from historical records that settlers in inland northern Sweden brought harvested seed to millers across the border (Kjellström, 2012). To what extent the traded seed was grown in Norway, or whether the Swedish farmers brought back seed from the other side of the border is not clear; but from the results in this study, it seems likely that seed trade across the borders of northern Fennoscandia had an impact on the genetic composition of the cultivated barley.

Historical sources mention that during times of crop failure, seed was transported in large quantities from Estonia to Sweden (Heckscher, 1935), but traces of such trade is not evident in this study. The extant accession from Estonia, VIR2143, clearly and consistently separates from the historical Fennoscandian accessions (Figures 3d and 4a). More samples with origins around the eastern Baltic, preferably from historical collections, are needed to clarify the relationship between the Estonian and Fennoscandian landrace barley.

A striking feature from the genetic structure is its apparent independence from national borders within the region, contrary to what historical records suggest. Not only in the north but also along the whole north–south range of Fennoscandia, we see genetic clustering shared between countries along the approximately same latitude. For example, in 1867–1868, just a few decades before the sampling of the accessions studied here, a severe crop failure in northern Sweden resulted in large quantities of barley for food and seed being imported from southern Sweden (Häger *et al.*, 1978). In spite of this, no mixing of south Swedish barley in the landraces from northern Sweden is evident. Although seeds were traded and documented human migrants presumably brought seeds with them, it is evident that long-term cultivation has been of the genotypes locally available, rather than imported seed. Possibly, adaptation to different climatic conditions has favored cultivation of local genotypes over imported ones. We are currently investigating the distribution of genetic diversity in candidate genes for climate adaptation to explore their role in barley cultivation.

DATA ARCHIVING

Genotype data available from the Dryad Digital Repository:doi:10.5061/dryad.293hd

ACKNOWLEDGEMENTS

This work was funded by the Swedish Research Council for Environment, Agricultural Sciences and Spatial Planning (FORMAS) and the Lagersberg Foundation. Seed material was kindly provided by Fredrik Ottosson at NordGen, Külli Annamaa at the Jögevaa Plant Breeding Institute, Igor Loskutov at the N.I. Vavilov Institute of Plant Industry, Annika Michelsson at Mustiala Agricultural College, Hannu Ahokas at MTT and Torbjørn Alm at Tromsø University Museum. Maria Lundström is acknowledged for skillful technical assistance.

REFERENCES

1. Avise JC (2009). Phylogeography: retrospect and prospect. *J Biogeogr* 36: 3–15.
2. Badr A, Muller K, Schafer-Pregl R, El Rabey H, Effgen S, Ibrahim HH *et al.* (2000). On the origin and domestication history of barley (Hordeum vulgare).*Mol Biol Evol* 17: 499–510.
3. Brown TA (1999). How ancient DNA may help in understanding the origin and spread of agriculture. *Philos Trans R Soc Lond B Biol Sci* 354: 89–98
4. Börner A, Chebotar S, Korzun V (2000). Molecular characterization of the genetic integrity of wheat (Triticum aestivum L.) germplasm after long-term maintenance. *Theor Appl Genet* 100: 494–497.
5. Caldwell KS, Russell J, Langridge P, Powell W (2006). Extreme population-dependent linkage disequilibrium in an inbreeding plant species, Hordeum vulgare. *Genetics* 172: 557–567.
6. Camacho Villa TC, Maxted N, Scholten M, Ford-Lloyd B (2005). Defining and identifying crop landraces. *Plant Genet Res* 3: 373–384.

7. Chebotar S, Roder MS, Korzun V, Borner A (2002). Genetic integrity of ex situ genebank collections. *Cell Mol Biol Lett* 7: 437–444.
8. Close TJ, Bhat PR, Lonardi S, Wu Y, Rostoks N, Ramsay L *et al.* (2009). Development and implementation of high-throughput SNP genotyping in barley. *BMC Genomics* 10: 582.
9. Conrad DF, Jakobsson M, Coop G, Wen X, Wall JD, Rosenberg NA *et al.* (2006). A worldwide survey of haplotype variation and linkage disequilibrium in the human genome. *Nat Genet* 38: 1251–1260.
10. De Carvalho MAP, Bebeli PJ, Bettencourt E, Costa G, Dias S, Dos Santos TM*et al.* (2013). Cereal landraces genetic resources in worldwide GeneBanks. A review. *Agron Sustain Dev* 33: 177–203.
11. ESRI (2011). *ArcGIS Desktop: Release 10.* Environmental Systems Research Institute: Redlands, CA.
12. Evanno G, Regnaut S, Goudet J (2005). Detecting the number of clusters of individuals using the software STRUCTURE: a simulation study. *Mol Ecol* 14: 2611–2620.
13. Excoffier L, Lischer HEL (2010). Arlequin suite ver 3.5: a new series of programs to perform population genetics analyses under Linux and Windows.*Mol Ecol Res* 10: 564–567.
14. Falush D, Stephens M, Pritchard JK (2003). Inference of population structure using multilocus genotype data: linked loci and correlated allele frequencies. *Genetics* 164: 1567–1587.
15. Flygare I (2011). The structure of agriculture. In: Jansson U, Wastenson L, Aspenberg P (eds.) *National atlas of Sweden. Agriculture and forestry in Sweden since 1900—a cartographic description.* Norstedt: Stockholm. p 58–70.
16. Fogelqvist J, Niittyvuopio A, Agren J, Savolainen O, Lascoux M (2010). Cryptic population genetic structure: the number of inferred clusters depends on sample size. *Mol Ecol Res* 10: 314–323.
17. Fuller DQ, Willcox G, Allaby RG (2011). Cultivation and domestication had multiple origins: arguments against the core area hypothesis for the origins of agriculture in the Near East. *World Archaeol* 43: 628–652.

18. Hagenblad J, Zie J, Leino MW (2012). Exploring the population genetics of genebank and historical landrace varieties. *Genet Res Crop Evol* 59: 1185–1199.
19. Heckscher EF (1935-1949). *Sveriges ekonomiska historia från Gustav Vasa*. Bonnier: Stockholm.
20. Herstad J (2000). *I helstatens grep: kornmonopolet.* Tano Aschehoug: Oslo, 1735–1788.
21. Hill WG, Robertson A (1968). Linkage disequilibrium in finite populations.*Theor Appl Genet* 38: 226–231
22. Hunt HV, Campana MG, Lawes MC, PARK YJ, Bower MA, Howe CJ *et al.* (2011). Genetic diversity and phylogeography of broomcorn millet (Panicum miliaceum L.) across Eurasia. *Mol Ecol* 20: 4756–4771.
23. Hutchison DW, Templeton AR (1999). Correlation of pairwise genetic and geographic distance measures: Inferring the relative influences of gene flow and drift on the distribution of genetic variability. *Evolution* 53: 1898–1914.
24. Häger O, Torell C, Villius H (1978). *Ett satans år: Norrland 1867.* Sveriges Radio: Stockholm.
25. Isaac AD, Muldoon M, Brown KA, Brown TA (2010). Genetic analysis of wheat landraces enables the location of the first agricultural sites in Italy to be identified. *J Archaeol Sci* 37: 950–956
26. Jakobsson M, Rosenberg NA (2007). CLUMPP: a cluster matching and permutation program for dealing with label switching and multimodality in analysis of population structure. *Bioinformatics* 23: 1801–1806.
27. Johnston SE, Lindqvist M, Niemela E, Orell P, Erkinaro J, Kent MP *et al.* (2013). Fish scales and SNP chips: SNP genotyping and allele frequency estimation in individual and pooled DNA from historical samples of Atlantic salmon (Salmo salar). *BMC Genom* 14: 439.
28. Jombart T, Devillard S, Balloux F (2010). Discriminant analysis of principal components: a new method for the analysis of genetically structured populations. *BMC Genet* 11: 1–15.
29. Jones G, Jones H, Charles MP, Jones MK, Colledge S, Leigh FJ *et al.* (2012). Phylogeographic analysis of barley DNA as evidence

for the spread of Neolithic agriculture through Europe. *J Archaeol Sci* 39: 3230–3238.

30. Jones H, Civan P, Cockram J, Leigh FJ, Smith LMJ, Jones MK *et al.* (2011). Evolutionary history of barley cultivation in Europe revealed by genetic analysis of extant landraces. *BMC Evol Biol* 11: 320.
31. Jones H, Lister DL, Bower MA, Leigh FJ, Smith LM, Jones MK (2008). Approaches and constraints of using existing landrace and extant plant material to understand agricultural spread in prehistory. *Plant Gen Res* 6: 98–112.
32. Kjellström R (2012). Nybyggarliv i Vilhelmina, Uppsala, Kungl. Gustav Adolfs Akademien för svensk folkkultur.
33. Lascoux M, Petit RJ (2010). The 'New Wave'in plant demographic inference: more loci and more individuals. *Mol Ecol* 9: 1075–1078.
34. Leino MW (2010). Frösamlingar på museum. *Nord Mus* 2010: 96–108.
35. Leino MW, Boström E, Hagenblad J (2013). Twentieth-century changes in the genetic composition of Swedish field pea metapopulations. *Heredity*110: 338–346.
36. Leino MW, Hagenblad J (2010). Nineteenth century seeds reveal the population genetics of landrace barley (Hordeum vulgare). *Mol Biol Evol* 27: 964–973.
37. Leino MW, Hagenblad J, Edqvist J, Strese EMK (2009). DNA preservation and utility of a historic seed collection. *Seed Sci Res* 19: 125–135.
38. Lister DL, Bower MA, Jones MK (2010). Herbarium specimens expand the geographical and temporal range of germplasm data in phylogeographic studies. *Taxon* 59: 1321–1323.
39. Lister DL, Jones H, Jones MK, O'Sullivan DM, Cockram J (2013). Analysis of DNA polymorphism in ancient barley herbarium material: Validation of the KASP SNP genotyping platform. *Taxon* 62: 779–789.
40. Lister DL, Thaw S, Bower MA, Jones H, Charles MP, Jones G *et al.* (2009). Latitudinal variation in a photoperiod response gene in European barley: insight into the dynamics of agricultural spread from 'historic' specimens. *J Archaeol Sci* 36: 1092–1098.

41. Londo JP, Chiang Y-C, Hung K-H, Chiang T-Y, Schaal BA (2006). Phylogeography of Asian wild rice, Oryza rufipogon, reveals multiple independent domestications of cultivated rice, Oryza sativa. *Proc Natl Acad Sci USA* 103: 9578–9583.

42. Lunden K (2004). Recession and new Expansion. In: Almås R (Ed) *Norwegian Agricultural History*. Tapir Academic Press: Trondheim, 141–232.

43. Malysheva-Otto LV, Ganal MW, Roder MS (2006). Analysis of molecular diversity, population structure and linkage disequilibrium in a worldwide survey of cultivated barley germplasm (Hordeum vulgare L.). *BMC Genet* 7: 6.

44. Moragues M, Comadran J, Waugh R, Milne I, Flavell AJ, Russell JR (2010). Effects of ascertainment bias and marker number on estimations of barley diversity from high-throughput SNP genotype data. *Theor Appl Genet* 120: 1525–1534.

45. Morrell PL, Clegg MT (2007). Genetic evidence for a second domestication of barley (Hordeum vulgare) east of the Fertile Crescent. *Proc Natl Acad Sci USA* 104: 3289–3294.

46. Morrell PL, Toleno DM, Lundy KE, Clegg MT (2005). Low levels of linkage disequilibrium in wild barley (Hordeum vulgare ssp. spontaneum) despite high rates of self-fertilization. *Proc Natl Acad Sci USA* 102: 2442–2447.

47. Nei M (1973). Analysis of gene diversity in subdivided populations. *Proc Natl Acad Sci USA* 70: 3321–3323.

48. Nielsen R, Hubisz MJ, Clark AG (2004). Reconstituting the frequency spectrum of ascertained single-nucleotide polymorphism data. *Genetics*168: 2373–2382.

49. Nordborg M, Hu TT, Ishino Y, Jhaveri J, Toomajian C, Zheng H *et al*. (2005). The pattern of polymorphism in Arabidopsis thaliana. *PLoS Biol* 3: e196.

50. Oliveira HR, Campana M, Jones H, Hunt H, Leigh F, Lister DL *et al*. (2012). Tetraploid wheat landraces in the Mediterranean basin: taxonomy, evolution and genetic diversity. *PLoS One* 7: e37063.

51. Oliveira HR, Hagenblad J, Leino MW, Leigh FJ, Lister DL, Penã-Chocarro L *et al*. (2014). Wheat in the Mediterranean revisited—tetraploid wheat landraces assessed with elite bread wheat single-nucleotide polymorphism markers. *BMC Genetics* 15: 54.

52. Olsen KM, Schaal BA (1999). Evidence on the origin of cassava: phylogeography of Manihot esculenta. *Proc Natl Acad Sci USA* 96: 5586–5591.
53. Osvald H (1959). *Åkerns nyttoväxter*. Stockholm: Sv. litteratur.
54. Palmer SA, Smith O, Allaby RG (2012). The blossoming of plant archaeogenetics. *Ann Anat* 194: 146–156.
55. Pandey M, Wagner C, Friedt W, Ordon F (2006). Genetic relatedness and population differentiation of Himalayan hulless barley (Hordeum vulgare L.) landraces inferred with SSRs. *Theor Appl Genet* 113: 715–729.
56. Parzies HK, Spoor W, Ennos RA (2000). 'Genetic diversity of barley landrace accessions (Hordeum vulgare ssp vulgare) conserved for different lengths of time in ex situ gene banks.'. *Heredity* 84: 476–486.
57. Pritchard JK, Stephens M, Donnelly P (2000). Inference of population structure using multilocus genotype data. *Genetics* 155: 945–959.
58. R Development Core Team (2013). *R: A language and environment for statistical computing*. R Foundation for Statistical Computing: Vienna, Austria.
59. Ren X, Nevo E, Sun D, Sun G (2013). 'Tibet as a potential domestication center of cultivated barley of China.'. *PLoS ONE* 8: 1.
60. Rodriguez M, Rau D, O'Sullivan D, Brown AHD, Papa R, Attene G (2012). Genetic structure and linkage disequilibrium in landrace populations of barley in Sardinia. *Theor Appl Genet* 125: 171–184.
61. Rosenberg NA (2004). DISTRUCT: a program for the graphical display of population structure. *Mol Ecol Notes* 4: 137–138.
62. Rostoks N, Mudie S, Cardle L, Russell J, Ramsay L, Booth A *et al*. (2005). Genome-wide SNP discovery and linkage analysis in barley based on genes responsive to abiotic stress. *Mol Genet Genomics* 274: 515–527.
63. Rostoks N, Ramsay L, MacKenzie K, Cardle L, Bhat PR, Roose ML *et al*. (2006). Recent history of artificial outcrossing facilitates whole-genome association mapping in elite inbred crop varieties. *Proc Natl Acad Sci USA*103: 18656–18661.

64. Roullier C, Benoit L, McKey DB, Lebot V (2013). Historical collections reveal patterns of diffusion of sweet potato in Oceania obscured by modern plant movements and recombination. *Proc Natl Acad Sci USA* 110: 2205–2210.

65. Russell J, Dawson IK, Flavell AJ, Steffenson B, Weltzien E, Booth A *et al.* (2011). Analysis of >1000 single nucleotide polymorphisms in geographically matched samples of landrace and wild barley indicates secondary contact and chromosome-level differences in diversity around domestication genes.*New Phytol* 191: 564–578.

66. Saisho D, Purugganan MD (2007). Molecular phylogeography of domesticated barley traces expansion of agriculture in the Old World.*Genetics* 177: 1765–1776.

67. Steiner AM, Ruckenbauer P, Goecke E (1997). Maintenance in genebanks, a case study: contaminations observed in the Nurnberg oats of 1831. *Gen Res Crop Evol* 44: 533–538.

68. Svensson EM, Anderung C, Baubliene J, Persson P, Malmström H, Smith C *et al.* (2007). Tracing genetic change over time using nuclear SNPs in ancient and modern cattle. *Anim Genet* 38: 378–383.

69. Van Heerwaarden J, Doebley J, Briggs WH, Glaubitz JC, Goodman MM, Gonzalez JJS *et al.* (2011). Genetic signals of origin, spread and introgression in a large sample of maize landraces. *Proc Natl Acad Sci USA*108: 1088–1092.

70. Weir BS, Cockerham CC (1984). Estimating F-statistics for the analysis of population structure. *Evolution* 38: 1358–1370.

71. Yahiaoui S, Igartua E, Moralejo M, Ramsay L, Molina-Cano JL, Ciudad FJ *et al.* (2007). Patterns of genetic and eco-geographical diversity in Spanish barleys. *Theor Appl Genet* 116: 271–282.

72. Zeven AC (1998). Landraces: A review of definitions and classifications.*Euphytica* 104: 127–139.

Chapter 5

Plant Phenotypic Plasticity in Response to Environmental Factors

Loretta Gratani

Department of Environmental Biology, Sapienza University of Rome, P.le A. Moro 5 00185, Rome, Italy

ABSTRACT

Plants are exposed to heterogeneity in the environment where new stress factors (i.e., climate change, land use change, and invasiveness) are introduced, and where inter- and intraspecies differences may reflect resource limitation and/or environmental stress factors. Phenotypic plasticity is considered one of the major means by which plants can cope with environmental factor variability. Nevertheless, the extent to which phenotypic plasticity may facilitate survival under environmental condition changes still remains largely unknown because results are sometimes controversial. Thus, it is important to identify plant functional traits in which plasticity may play a determinant role in plant response to global change as well as on the ecological consequences at an ecosystem level for the competition between wild and invasive species, considering that species with a greater adaptive plasticity may be more likely to survive in novel environmental conditions. In the near future, it will be important to increase long-term studies on natural populations in order to understand plant response to environmental factor fluctuations including climate change. There is the necessity to analyze variations at phenotypic and genetic levels for the same species and, in particular, for endemic and rare species because these could have drastic effects at an ecosystem level.

INTRODUCTION

Literature on phenotypic plasticity has increased expanding from the initial focus on abiotic factors to that of biotic ones [1–3] and, in recent years, taking into consideration plant response to global climate change, land use change and plant invasiveness [4, 5] (Table 1). Thus, fundamental questions for evolutionary ecologists in a global change context are how plant species will respond to these new scenarios and what mechanisms will be involved in the process [6, 7]. The understanding of phenotypic plasticity will be crucial for predicting changes in species distribution, community composition, and crop productivity under global change conditions [8, 9]. Nevertheless, the theme of phenotypic plasticity is complex and researchers do not always arrive at the same conclusions and results are sometimes controversial.

Table 1: List of the mentioned species and corresponding references

Species	References
Abies alba	[27]
Acer pseudoplatanus	[27]
Acer saccharum	[28]
Betula papyrifera	[28]
Campanula thyrsoides	[29]

Coffea arabica	[30, 31]
Corylus avellana	[32–34]
Crepis pygmaea subsp. pygmaea	[35]
Cyclobalanopsis multinervis	[36]
Epilobium fleischeri	[29]
Fagus sylvatica	[27, 37–39]
Fraxinus excelsior	[32–34]
Geum reptans	[29]
Glycine max	[40]
Glycine soja	[40]
Hedera helix	[41]
Ilex aquifolium	[42]
Isatis apennina	[35]
Lythrum salicaria	[43, 44]
Myrtus communis	[45]
Ostrya virginiana	[28]
Phillyrea latifolia	[41, 46]
Picea abies	[27, 47]
Pinus halepensis	[48–50]
Pinus nigra	[48]
Pinus pinaster	[48]
Pinus pinea	[48]
Pinus sylvestris	[48, 51]
Pinus uncinata	[48]
Pistacia lentiscus	[46, 52–55]
Poa alpina	[29]
Populus tremula	[32, 33]
Populus tremuloides	[28]
Quercus aliena var. acutiserrata	[36]
Quercus coccifera	[50, 56]
Quercus faginea	[48]
Quercus ilex	[19, 46, 48, 55, 57–61]
Quercus mongolica var. crispula	[62, 63]
Quercus petraea	[48]
Quercus pyrenaica	[48]

Quercus robur	[48]
Quercus suber	[48]
Rhododendron ponticum	[42]
Rubia peregrina	[41, 64]
Ruscus aculeatus	[41]
Sesleria nitida	[65]
Shorea disticha	[66]
Shorea trapezifolia	[66]
Shorea worthingtonii	[66]
Smilax aspera	[41]
Stellaria longipes	[67]
Taraxacum officinale	[68]
Tilia cordata	[32–34]
Viburnum tinus	[41]

Phenotypic plasticity has been defined as a change in the phenotype expressed by a single genotype in different environments. Bradshaw [10] recognized that phenotypic plasticity could itself be under genetic control and therefore subjected to selective pressure. Scheiner and Goodnight [11] show that there is no reason to believe that the selection of plastic and genetic variations need necessarily be coupled. Nevertheless, a population could respond to an extremely variable environment by becoming both more plastic and more genetically variable. Literature suggests that phenotypic plasticity can evolve when there is a sufficient genetic variation [12, 13] due to genetic correlations with other traits that are under selection or to genetic drift [14]. Since phenotypic plasticity influences environmental tolerance, different plastic responses may contribute to differences in the range of environments that species inhabit [15]. In particular, the environment can induce changes in the individual's behavior at a morphological and/or physiological level [16] and such changes may be crucial to survival in heterogeneous and variable conditions [17–20]. For certain morphological traits, phenotypic plasticity has been shown to reflect genetic correlations relatively well, and traits belonging to the same suite of characters are more highly genetically and phenotypically correlated than traits from different suites [21]. The selection for photosynthetic traits may often operate indirectly via correlation with other traits, emphasizing the importance of viewing the phenotype as an integrated function of growth, morphology, life history, and physiology [22]. The timing of plant development can itself be plastic [23] and many phenotypic responses to environmental stress factors may be the consequence of growth reduction due to resource limitations [24, 25]. Differences among species and populations may reflect different selective pressures on plasticity, different limitations acting upon the maximization of plasticity, or a combination of both [26]. The potential plastic response of a given trait may be large but the observed plasticity may be lowered by resource limitations or environmental stress factors [14].

The particular way by which a genotype varies in its expression across a range of environments can be described by a reaction norm which is genetically determined [69]. The reaction norm for any specific trait of a genotype can be visualized as a line or a curve on a two-dimensional plot of the environmental value versus the phenotypic value (Figure 1). Phenotypic plasticity can be visualized as a change in the slope of the reaction norm between ancestral and derived populations or species [70, 71]. Such change has been shown to occur in nature between species subjected to different selection pressures [72–74]. Plasticity is what makes the appearance of an environmentally induced novel phenotype possible, and a process of selection on the expression of such phenotype in a new environment may end up "fixing" (genetically assimilating) it by altering the shape of the reaction norm [75]. Thus, plasticity could facilitate the expression of relatively well-adapted phenotypes under novel conditions (e.g., after migration to new geographical areas) improving the performance of the population and resulting in the genetic assimilation of the trait in the new environment. This has the potential to explain a variety of evolutionary ecological processes [14, 75].

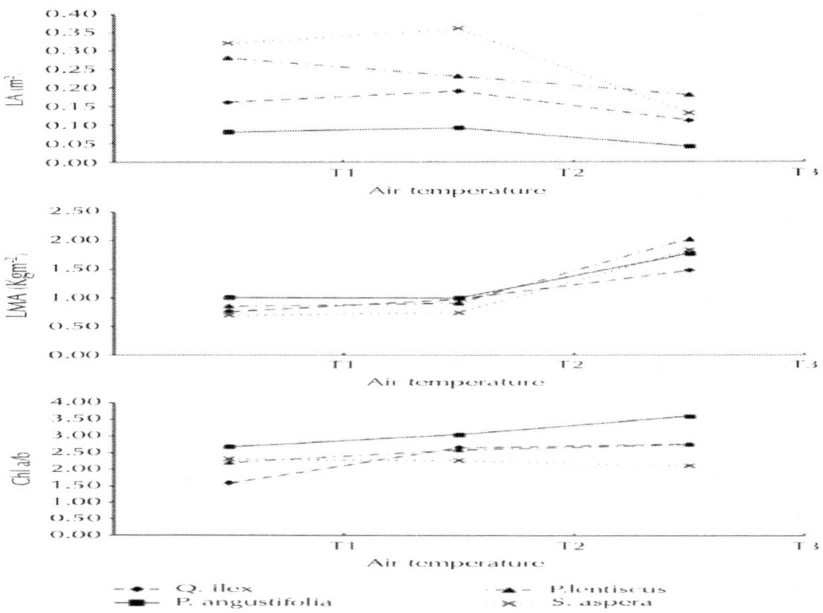

Figure 1: The reaction norm for any specific traits of a genotype can be visualized as a line on a plot of the environmental value versus the phenotypic value. In this example, the response of four evergreen species (Quercus ilex, Phillyrea angustifolia, Pistacia lentiscus, and Smilax aspera) in three different sites inside the Castelporziano Estate (Latium coast near Rome, Italy) to a gradient of air temperature (T1 = 16.5°C, T2 = 17.2°C, and T3 = 21.9°C) is shown [55]. Lines represent the species and slope the phenotypic response. Among the cooccurring species, S. aspera, growing in the understory of the forest, has a larger morphological than a physiological plasticity. Q. ilex is the species with the largest morphological and physiological plasticity. Leaf area (LA), leaf mass area (LMA), and chlorophyll a to b ratio (Chl a/b) are shown.

Indexes can facilitate comparison of different studies [76, 77], set of species or populations within a given species by considering experimental data in research on plasticity [19, 56, 57]. However, at least 17 different indexes have been employed as a measure of the phenotypic plasticity but can be flawed and applied in different ways. Most of them cannot be standardized across traits or compared among different species [78] complicating comparison among studies [3, 79]. Moreover, measures of phenotypic plasticity are strongly related to the context and may not be comparable across different studies where different gradients and/or species have been used [77]. Since information about plasticity is structured in a way that makes it difficult for quantitative and comparative analysis, Poorter et al. [79] proposed a method to fill this gap by building a large database which currently contains data on 1000 experiments and 800 species. This approach could serve as a benchmark for future phenotyping efforts as well as for modelling global change effects on both wild species and crops [79].

PHYSIOLOGICAL, ANATOMICAL, AND MORPHOLOGICAL PLASTICITY

Physiological, morphological, and anatomical plasticity may have a different role in plant adaption to environmental changes. In particular, plasticity for physiological and life-history traits may allow plants to grow and reproduce in spatially or temporally variable environments [46, 80]. Physiological plasticity is more linked to an enhanced capacity to colonize gaps and open areas [81, 82] because it ensures adjustments of gas exchange in response to environmental factor changes in a short term. This aspect evidences the importance of physiological plasticity in plant acclimatization to adverse environments where morphological and anatomical plasticity play a secondary role [83]. In fact, plants growing in stress conditions tend to have a conservative leaf morphological pattern to avoid the production of structures too expensive to be sustained [84, 85]. Moreover, morphological plasticity is more linked to an enhanced plant capacity to grow in forest understories [37, 82] by having an important role in resource acquisition [64, 86].

PLANT RESPONSE TO LIGHT

The heterogeneous light environment within a plant canopy brings about different stress factors for leaves in different canopy positions. One of the expressions of plant phenotypic plasticity is the modification of leaf traits to the light gradient [87] and the reduced red/far red ratio [2] from the top to the bottom of the tree canopy, mainly during leaf formation [69]. One of the main morphological traits which changes in response to light variations is the specific leaf area (SLA, ratio between leaf area and leaf dry mass). The plasticity of SLA implies the morphogenetic control of leaves which tends to increase leaf area in the shade in order to intercept more light while on the contrary there is a genetic or physiological limitation to the total leaf volume as well as a resource limitation [88]. SLA reflects leaf thickness [89] and the relative proportion of assimilatory, conductive, and mechanical tissues [90]. In particular, the increased total lamina thickness in sun leaves compared to shade leaves is mainly due to the greater palisade parenchyma, spongy parenchyma, and epidermal tissue thickness, suggesting that leaf internal structure may play an important role in light capture [91]. In a research on leaf morphology of woody angiosperms, Wylie [92] shows that, as a result of shading, leaf thickness decreased, on an average, by 54%, the volume of the palisade parenchyma by 60%, and the epidermal thickness by 17%. McClendon [93] and Gratani et al. [46] showed a strong correlation between leaf thickness and the light-saturated rate of photosynthesis per unit of leaf area.

Sun leaves with respect to shade leaves, on an average, have a higher photosynthetic rate on a leaf area basis which is associated to a higher chlorophyll (Chl) a/b ratio, a significantly lower light-harvesting Chl a/bprotein (LHCII), a lower stacking degree of thylakoids [94, 95], and a higher nitrogen (N) content per unit of leaf area [46, 96, 97] since approximately half of N is invested in the photosynthetic proteins [96]. Hirose and Werger [98] suggest that N varies with light availability in plant canopies in such a way as to optimize the daily canopy photosynthesis. Thus, the N investment is related to light in that more N is allocated to carboxylation in sun leaves and to light harvesting in shade leaves with variations in the photosynthetic capacity [99] (Figure 2). The photosynthetic capacity and N content are generally both higher in leaves under high light conditions [96, 98, 100]. The higher photosynthetic rates of sun leaves are supported by higher stomatal conductance and stomatal density to maximize CO_2 absorption [101]. On the contrary, leaf trait adjustments to low light increase the capacity of light absorption at the expense of the photosynthetic capacity minimizing carbon loss through respiration [96].

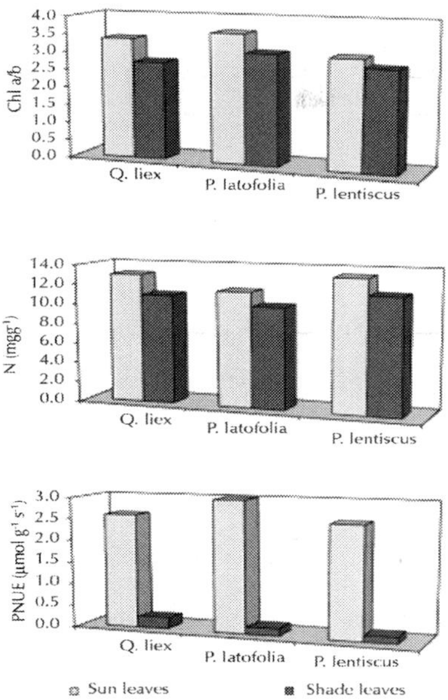

Plasticity index	Q. liex	P. latofolia	P. lentiscus
Morphological traits			
DM	0.16	0.11	0.14
LA	0.50	0.38	0.32
LMA	0.41	0.28	0.24
SLA	0.41	0.29	0.24
Mean value	0.37	0.26	0.23
Anatomical trails			
Leaf thickness	0.41	0.30	0.20
Palisade parenchyma thickness	0.47	0.37	0.28
Spongy Parenchyma thickness	0.41	0.25	0.12
Adaxial cuticle thickness	0.19	0.31	0.28
Abaxial cuticle thickness	0.31	0.33	0.27
Adaxial epidermis thickness	0.27	0.28	0.18
Abaxial epidermis thickness	0.14	0.05	0.01
Stomatal density	0.28	0.20	0.08
Guard cell length	0.08	0.12	0.00
Mean value	0.28	0.25	0.16
Biochemical tails			
N	0.15	0.12	0.12
Chl a + b	0.35	0.40	0.47
Chl/Car	0.32	0.32	0.42
Chl a/b	0.16	0.15	0.08
Chl/N	0.44	0.43	0.51
Mean value	0.28	0.28	0.32
Physiological trails			
P_N	0.96	0.97	0.95
g_s	0.67	0.56	0.59
WUE	0.86	0.92	0.85
PNUE	0.92	0.95	0.93
Ψ_m	0.14	0.12	0.10
Mean value	0.71	0.70	0.69

Figure 2: Leaf trait variations in response to the light gradient from the top to the bottom of the canopy of three evergreen shrubs (Pistacia lentiscus, Phillyrea latifolia, and Quercus ilex) cooccurring in the Mediterranean maquis [46]. All the considered species have significantly thicker sun leaves compared to shade leaves (on an average 0.45 times) due to the palisade parenchyma (61%, mean value), spongy parenchyma (38%, mean value), and the adaxial cuticle (36%, mean value) thickness. The higher leaf consistency (i.e., higher leaf mass area, LMA) of sun leaves can be used as a measure of investment per unit of leaf area in conditions of full sun. Moreover, shade leaves have a lower chlorophyll a to b ratio (Chl a/b) (13%, mean of the considered species) due to the higher chlorophyll b content (89%, mean of the considered species) since it is usually the main component of the LHCP (light-harvesting protein), which is higher in the shade conditions. Sun leaves allocate, on an average, 16% higher nitrogen (N) content than shade leaves reflecting an increase in carboxylating enzymes (RUBISCO) and proteins, responsible for the photosynthetic electron transport in full sun. The phenotypic plasticity of the considered species is higher for leaf physiological traits (0.86), and among them, net photosynthesis () and the photosynthetic nitrogen use efficiency (PNUE) have a larger plasticity (0.96 and 0.93, resp.). The largest phenotypic plasticity ofQ. ilex (0.41) among the cooccurring species (0.36, mean value) reflects its wider ecological distribution area. Leaf

dry mass (DM), leaf area (LA), specific leaf area (SLA), total chlorophyll a + b content (Chl a +b), chlorophyll to carotenoid ratio (Chl/Car), chlorophyll to nitrogen ratio (Chl/N), net photosynthesis (), stomatal conductance (), water use efficiency (WUE), and leaf water potential at midday () are shown.

Changes in the efficiency of light interception and in the costs for light harvesting along the light gradient from the top to the bottom of the plant canopy are the major means by which an efficient light harvesting is achieved. In particular, at shoot scale, foliage inclination angle and foliage spatial aggregation are the major determinants of light harvesting while, at the canopy scale branching frequency, foliage distribution and biomass allocation to leaves modify light harvesting significantly [102]. Intracanopy plasticity has important impacts on many aspects of tree biology, potentially contributing to the whole-canopy performance via effects of light penetration through the canopy and on the energy, carbon, and water balance of the individual leaves [97, 103, 104]. Plasticity within the same plant, in a temporary heterogeneous environment, may play an adaptive role in strong seasonal climates such as the Mediterranean climate [20].

Moreover, leaf trait variations in response to the light gradient within the canopy change in different species and forest types [105–108]. In forests with dense foliage, the upper layers absorb the majority of the incoming radiation. In broadleaf and conifer forests of the temperate zone, on an average, 3–10% of the incident photon flux density (PFD) penetrates the tree canopy [69]. Gratani and Foti [109] show that SLA increases, on an average, by 21% from the dominant tree layer of a mixed deciduous broadleaf forest to the dominated layer in response to a decrease in the light level. The same trend is measured from the dominant tree layer of a Mediterranean evergreen forest to the dominated layer by the 54% SLA increase, associated with a 9.5% lower chlorophyll a/b ratio and 86% lower photosynthetic rate [110]. Mendes et al. [45] analyzed the large morphological and anatomical leaf trait variations to different light conditions for Myrtus communis, an evergreen sclerophyllous species which grows in Mediterranean areas from open clearings to understories, with a 26% increase of the SLA from sun to shade conditions. Nevertheless, taxonomically different species cooccurring in the same habitat often share common morphological and physiological traits, reflecting a convergent evolution in response to environmental factors [111]. The large plasticity in the structure of the mesophyll in concert with other traits seems to enable these species to cope with different environmental regimes (i.e., Mediterranean type climate), attaining a wider habitats range [111].

SHADE TOLERANCE

Shade tolerance is one of the most important ecological factors with respect to interspecific competition among forest trees of the temperate climate zone. The shade tolerance of a given plant is defined as the minimum light under which a plant can survive; however, only a fraction of plants can reproduce under these conditions. Thus, a biological definition of shade tolerance must consider the whole life cycle of the plant from early survival and growth to reproduction [112]. In general, sun and shade leaves are thinner in tolerant species than in intolerant species [113]. The overall trend in literature suggests that early successional and light demanding species are more plastic than late-successional and shade-tolerant species [114–116]. Nevertheless, there is increasing evidence that indicates that adjustments are not necessarily related to the successional status of the species [117, 118] (Table 2).

Table 2: Phenotypic plasticity of morphological and physiological leaf traits for the mentioned shade-tolerant, shade-intolerant, and intermediate species and corresponding references

	Species	Morphological plasticity	Physiological plasticity	References
Shade-tolerant	Fagus sylvatica	0.46	0.39	[37]
	Acer pseudo platanus	0.31	0.34	[27]
	Ilex aquifolium	0.54	0.24	[119]
	Quercus ilex	0.33	0.5	[46]
	Acer saccharum	0.3	0.3	[108]

Inter mediate	Acer rubrum	0.41	0.26	[108]
	Betula alleghaniensis	0.29	0.32	[108]
	Picea abies	0.13	0.22	[27]
	Quercus rubra	0.41	0.27	[108]
Shade-intolerant	Quercus robur	0.24	0.6	[37]
	Cistus incanus	0.18	0.38	[120]
	Abies procera	0.29	0.32	[108]

Numerous studies focus on acclimation of the photosynthetic properties of plant species to different light conditions; nevertheless, results sometime disagree. For example, Coffea arabica is an evergreen perennial tree from the African forest understory which is considered an obligatory shade species. Matos et al. [30] show that plasticity of physiological and biochemical traits in C. arabica is more important for acclimation to intracanopy light variations than morphological and anatomical trait plasticity. Nevertheless, Araujo et al. [31] show that the apparent inability of C. arabica shade leaves to optimize carbon gain under low light and the successful ability of sun leaves to prevent photoinhibition of photosynthesis under high light conditions are consistent with the findings that it performs well at full sun exposure. Niinemets et al. [32] showed that in temperate deciduous forests species differ in their shade tolerance with Corylus avellana having the largest tolerance, followed by Tilia cordata, Fraxinus excelsior, and Populus tremula [33, 34]. Wyka et al. [27] by comparing seedlings of two conifers (the shade-tolerant Abies alba and the moderately shade-tolerant Picea abies) and two deciduous angiosperms (the highly shade-tolerant Fagus sylvatica and the moderately shade-tolerant Acer pseudoplatanus) in two-year studies, tested the hypothesis that light-induced plasticity of physiological and anatomical traits was lower in highly shade-tolerant species than in moderate shade-tolerant ones. The authors concluded that shade tolerance was not a consistent predictor of anatomical plasticity and that, for many traits, differences between the light treatments were magnified in the second year, showing that anatomical and physiological adjustment to shade is a long-term process. Valladares et al. [37] investigated phenotypic plasticity of two species differing in their shade tolerance (the most shade-tolerant Fagus sylvaticaand the more light-demanding Quercus robur) hypothizing a lower plasticity in the species of deep shade than in the more light-demanding species, according to comparative studies of others woody plants [56, 85, 105]. Moreover, the authors suggested that an alternative hypothesis was also feasible (i.e., that shade tolerance can be achieved by enhanced plasticity) as argued previously for the same two species [121]. The results showed a mean phenotypic plasticity similar to the two species (Q. robur and F. sylvatica) due to the fact that while Q. robur has a higher physiological plasticity, F. sylvatica has a higher morphological plasticity, with one group of traits counteracting the effect of the other group on the mean plasticity index [37].

Differences in forest structure determine changes in the light gradient and, consequently, in the presence of shade-tolerant and intolerant species. In particular, the deciduous forests may consist of an upper layer with shade-intolerant tree species and a lower layer with shade-tolerant tree species [32], while in mixed forests, deciduous and evergreen species with contrasting ecological potentials can occupy different layers [122]. Poorter et al. [48] analyzed shade tolerance in tree species growing in some forests of Continental Spain showing that the wider canopy, lower height, and flat leaves of the Fagaceae (Quercus faginea, Quercus ilex,Quercus petraea, Quercus pyrenaica, Quercus robur, and Quercus suber) enhanced their shade tolerance compared to Pinaceae (Pinus halepensis, Pinus nigra, Pinus pinaster, Pinus pinea, Pinus sylvestris, and Pinus uncinata). On the contrary, resistance to fire [123], drought tolerance [124], and allometric plasticity of the shade-intolerant pines increased their rapid occupation of open spaces [125] and could be relevant to explain species ecological dominance. The shade-intolerant Populus tremuloides and Betula papyrifera, common species of the broad ecotone between the boreal and cold-temperate forests of central and eastern North America, possess traits that maximize whole plant carbon dioxide (CO_2) exchange leading to long-term growth and survival disadvantages in very low light, while the shade-tolerant Acer saccharum and Ostrya virginiana minimize CO_2 losses in low light at the expense of maximizing CO_2 gains in high light [28]. The same authors conclude that the shade-intolerant species have a larger plasticity than the shade-tolerant species. Ke and Werger [36] showed that the deciduous Quercus aliena var. acutiserrata is a gap-dependent species having advantage over the evergreen Cyclobalanopsis multinervis in forest gaps and clearings by the greater morphological and physiological plasticity of seedlings. The ability to acclimate to contrasting light environments is particularly important for tropical woody species because tree seedlings grown in the forest understory are prone to experience photoinhibition under an increased light level through formation of a gap [126]. High irradiance in tropical latitudes can cause chronic photoinhibition through impairment of photosystem II (PSII) reaction centres in the leaves of plants that have experienced gap openings [127, 128]. Nevertheless, species vary in their shade tolerance to the increased light after a gap creation. For example, in deciduous broadleaf forests of northern Japan, Quercus mongolica var. crispula is a gap-dependent species because its seedlings could survive at most a dozen years under a closed canopy even though they can germinate in low light environments [62, 63]. Koike et al. [129] in a research on early successional species, gap phase species, mid-successional species, and late successional species showed that acclimation potential of the species was expressed by their plasticity which was closely linked to the photosynthetic capability [130].

Despite the assumption that shade leaves develop in response to reduced light, other factors may also be involved, such as temperature and water stress [113]. It has been hypothesized that plants cannot tolerate combined shade and drought, as a result of a morphological trade-off [49, 131]. Nevertheless, Sack et al. [41] investigated six species (Phillyrea latifolia, Viburnum tinus, Rubia peregrina, Ruscus aculeatus, Hedera helix, and Smilax aspera) that cope with strong summer drought in the understory of mixed Quercus forests in southern Spain. All the species persisted in the shade (ca 3% daylight) and converged in features that conferred tolerance to shade plus drought by reducing the demand for resources. In particular, demand for water was reduced through a moderate-to-high below-ground mass fraction and low-to-moderate SLA, while demand for both low irradiance and water was reduced through a low-to-moderate leaf N content and leaves with a long life span. On an average, SLA is the trait most strongly correlated to shade tolerance [132].

Generalist species that grow in a range of moisture and light conditions within a forest, on an average, have a larger morphological plasticity than specialist species [66]. In particular, the more generalist Shorea distichaoccurring on most sites within the evergreen forests and extending over most areas of the South and Southeast Asia has a greater plasticity than Shorea trapezifolia, which is restricted to the lower slopes of the valley within the forest, and Shorea worthingtonii which is restricted to the ridge-tops.

ECOTYPES

Long-term selection can lead to the development of morphological and physiological adaptations to the local environment generating ecotypic differentiation in functional traits [133, 134]. Genotypes adapted to local environmental conditions are referred to as ecotypes [135]. When environments within the distribution area of a species differ, it is unlikely that any single phenotype confers high fitness in all situations. The distinction between phenotypic plasticity and local adaptation of an ecotype is based primarily upon genetic analysis and transplantation experiments [52]. In particular, spatial genetic differentiation along climatic gradients has been documented for many species [136–138] as well as for ecotype formations [52, 136, 138]. For example, ecotypes of Pinus taeda [139], Picea abies [47], Pinus sylvestris [51], Fagus sylvatica [38, 39], and Quercus coccifera [50] have adaptive features which are probably driven by the climate of the locality from which they originate. Moreover, species with extensive geographical ranges have the potential to exhibit a larger intraspecific variation in physiology, morphology, phenology, and growth rate [140]. Gratani et al. [19], Bonito et al. [57], and Pesoli et al. [58] compared plant and leaf trait plasticity in Quercus ilex seedlings from different provenances in Italy and grown in a common garden (Figure 3). Quercus ilex is a deep-rooted evergreen species widely distributed in the Mediterranean Basin [59] extending 6,000 km longitudinally from Morocco to France. This species seems to be limited to the southern range due to increased summer drought, and in altitude (it is distributed from the sea level to 1,100 m a.s.l.) by factors associated with low temperature [60]. The results show that seedlings from the more xeric provenance have a higher tolerance to drought stress by a higher leaf mass area (LMA) limiting transpiration and a higher stomata sensitivity to changes in water potential allowing a higher relative leaf water content (RWC). On the contrary, the reduced leaf area appears to be the best adaptive trait in response to winter cold stress at the northern limit of the distribution area, while the largest shoot production reflects the more favorable climatic conditions at the centre of the distribution area. Thus, while the favorable environmental conditions increase the phenotypic plasticity of Q. ilex morphological and physiological traits, the less favorable conditions (i.e., cold and drought stress) allow specialization. Michaud et al. [61] observed a homogeneous genetic structure of Q. ilex in the Mediterranean region with only slight geographic variations due to isolation (i.e., North Africa and Sicily) which supports the hypothesis that Q. ilex corresponds to a single genetic entity. According to these hypotheses, Balaguer et al. [56] show that phenotypic plasticity varies across the geographical range of Quercus coccifera, and among populations, suggesting that Q. coccifera ecotypic differentiation accounts for its occurrence in contrasting habitats. Wen et al. [40] studied the origin and evolution of cultivated soybean. They investigated genetic diversity, geographic differentiation, and genetic relationship among geographic ecotypes of cultivated (Glycine max) and wild (G. soja) soybeans growing in South-Central China, South-West China, and South China. The results showed that the wild accessions had relatively small genetic distances with all cultivated accessions and the Middle and Lower Changjiang valleys wild ecotypes were smaller compared to other wild ones, including their local wild counterparts. Therefore, it is inferred that the wild ancestors in southern China, especially those from Middle and Lower Changjiang valleys, might be the common ancestors of all cultivated soybeans. Nevertheless, exploitation of new habitats may be associated with the loss of plasticity and evolution of specialization [16]. Adaptation of species to geographic environmental variations often depends on genetic variations among seed sources [19, 141]. Nahum et al. [52] showed that Pistacia lentiscusecotypes growing in diverse habitats along a climatic gradient in Israel do not have any pattern of ecologically related genetic differentiation, and morphological and physiological differences are probably due to phenotypic plasticity. Thus, adaptive plasticity can expand environmental tolerance contributing to a wide distribution [53] of P. lentiscus around the Mediterranean region [52, 54]. Emery et al. [67] show that ecotypes of Stellaria longipes, an herbaceous perennial species growing along an altitudinal gradient on Plateau Mountain (Alberta) from the alpine tundra (i.e., higher altitude), has a lower plasticity than the ecotype from the prairie (i.e., lower altitude).

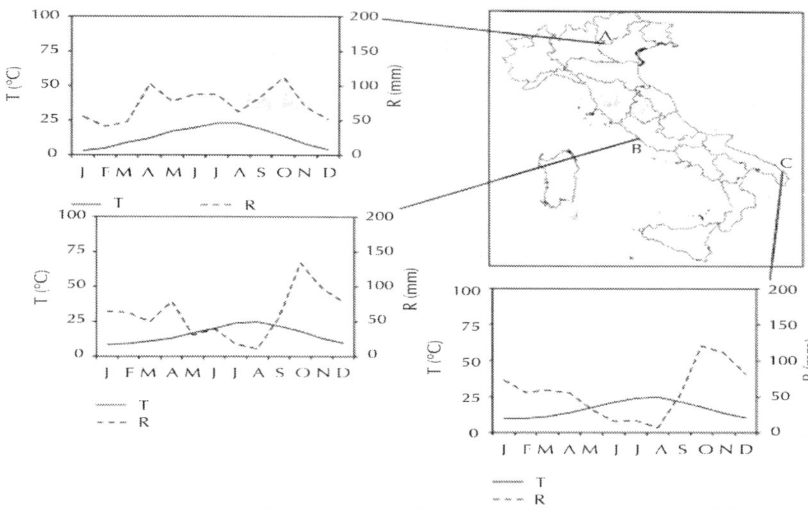

Plasticity index	A	B	C
Physiological traits			
P_N	0.73	0.73	0.67
g_s	0.73	0.79	0.77
E	0.84	0.84	0.82
WUE	0.88	0.87	0.82
Ψ_m	0.89	0.93	0.93
Morphological traits			
LA	0.24	0.27	0.33
DM	0.19	0.40	0.32
SLA	0.15	0.27	0.15
LMA	0.15	0.27	0.15
LTD	0.21	0.32	0.15
Mean value	0.50	0.57	0.51

Figure 3: Physiological, morphological, and anatomical leaf trait variations of Quercus ilex ecotypes from different provenances in Italy, grown from seeds collected in the native environments. Seedlings from the more xeric provenance (Frassanito, 40°13′N, 18°26′E, at sea level, site C) have a larger tolerance to drought by a larger leaf mass area (LMA) and leaf tissue density (LTD) and a higher stomatal sensitivity to changes in leaf water potential at predawn () contributing to a better water use efficiency (WUE) than the other seedlings [58]. In particular, the reduced leaf area (LA) appears to be the best adaptive trait in response to the winter stress at the northern distribution limit (Nago, 45°55′N, 10°53′E, 260 m a.s.l., site A) while the larger shoot and leaves production of Castelporziano seedlings reflects the favorable climatic condition of this locality (41°45′N, 12°26′E, at sea level, site B). The favorable environmental conditions at Castelporziano are

expressed throughout a larger psshenotypic plasticity while the cold stress at Nago and the drought stress at Frassanito allow specialization. Moreover, Q. ilexphenotypic plasticity is larger for physiological than for morphological and anatomical traits due to the capability of this species to grow in full sun as well as in shade conditions and colonize successfully new areas after fire by vegetative regeneration [142]. Frassanito shrubs by their larger tolerance to high temperature and limited water availability might have an advantage in response to the forecasted increase of air temperature and drought in the Mediterranean Basin. Net photosynthesis (), stomatal conductance (), leaf transpiration (E), and specific leaf area (SLA) are shown.

To accurately determine patterns of plasticity and investigate the ecological and evolutionary implications, it is important to better understand the environmental context in which phenotypes are expressed [2].

HIGH ALTITUDE MOUNTAIN PLANTS

The impact of global warming on terrestrial ecosystems has been shown to be greater in arctic tundra and high mountain regions than in low latitude areas [143]. In particular, the projected rate of global warming in mountain ecosystems is expected to be up to three times higher than the global average rate of warming recorded during the 20th century [144]. The biodiversity scenarios for the 21st century forecasts the reduction of alpine habitat and loss of many European high-mountain plants [145, 146]. Dirnböck et al. [147] hypothesize a rapid increase in plant species extinction risk. Moreover, species-specific reduction in fitness and diversity could change community dynamics by altering species competitive abilities. Recent studies [35] indicate substantial adaptive potential as reflected by high heritability estimates for traits likely to be selected [148]. Nevertheless, there is little information on the adaptive potential in environments that are particularly threatened by climate change such as high altitude mountain areas. One of the predicted consequences of global climate change is the movement of plant species to higher elevations and latitudes as the climate to which they are adapted is displaced [149]. A drastic decrease of the distribution area or even extinction of plant species can be the consequence of migration processes towards higher altitudes. Rates and patterns of these dynamics will be highly dependent on the habitat preference of a particular species and on its functional traits [150]. Steep environmental gradients and patchy distribution of habitats lead to small size and spatial isolation of populations and restrict gene flow [29]. It has been hypothesized that plant species populations may persist in their current areas and withstand environmental changes if they have a large adaptive capacity [151]. Gratani et al. [35], in a comparative study on two populations of Crepis pygmaea subsp. pygmaea andIsatis apennina growing at different altitudes on the Gran Sasso massif (Italy), addressed the question whether plasticity of morphological and physiological traits could be indicative of their future adaptive potential to survive global warming. The results underline that C. pygmaea has a significantly higher plasticity of both physiological and morphological traits than I. apennina. Thus, the hypothesized air temperature increase could drive C. pygmaea and I. apennina to higher altitudes in the Gran Sasso massif with C. pygmaea being favored by the highest plasticity. Studies on Sesleria nitida, a perennial herbaceous species growing at different altitudes along a narrow altitudinal gradient in the Central Apennines (Monte Terminillo, Italy), show that the lower photosynthetic rates at the higher elevation are justified by the stronger wind action and the lower soil water content of this site, while the lower SLA (i.e., the highest leaf mass area, LMA) contributes to limit leaf transpiration [65]. The extent to which SLA is phenotypically plastic or genetically fixed has important implications for the survival of populations under environmental condition changes [152]. Moreover, the results show a larger plasticity for physiological than for morphological and anatomical leaf traits of S. nitida the first being more useful under strong stress conditions which change in a short-term (Table 3). The relatively large phenotypic plasticity of S. nitida reflects its capability to maintain function under different environmental stress conditions and sustain the air temperature increase through a potential shift toward higher elevations. Stöcklin et al. [29] studied the consequences for the alpine landscape due to the evolutionary processes in four typical alpine plant species (Epilobium fleischeri, Geum reptans, Campanula thyrsoides, and Poa alpina). The within-population genetic diversity of the four species is large and mostly not related to altitude and population size. Nevertheless, genetic differentiation among populations is strongly increased with distance, thus suggesting considerable genetic drift among populations of alpine plants. Phenotypic variability is shaped by adaptive as well as by random evolutionary processes, and plastic responses to growth conditions seem to be crucial for survival of plants in the alpine landscape.

Table 3: Plasticity index (PI) of morphological and physiological leaf traits for Crepis pygmaea subsp. pygmaea, Isatis apennina [35], and Sesleria nitida [65]

	Morphological traits	PI	Physiological traits	PI
Crepis pygmaea subsp pygmaea	LA	0.20	PN	0.63
	DM	0.39	gs	0.65
	LMA	0.38	R	0.53
	LTD	0.42	WUE	0.36
			Ψ_m	0.73
			Chl	0.31
			RWC	0.21
			Pq	0.55
	Mean value	0.35	Mean value	0.50
Isatis apennina	LA	0.25	PN	0.42
	DM	0.32	gs	0.48
	LMA	0.30	R	0.72
	LTD	0.33	WUE	0.26
			Ψ_m	0.48
			Chl	0.17
			RWC	0.15
			Pq	0.37
	Mean value	0.30	Mean value	0.38

Sesleria nitida	Leaf thickness	0.24	PN	0.18
	Height of the major lateral vascular bundle	0.30	gs	0.39
	Width of the major lateral vascular bundle	0.24	E	0.33
	Height of the major lateral vascular bundle	0.39	R	0.37
	Width of the central vascular bundle	0.32	R/ PN	0.47
	Diameter of the xylematic vessels	0.18		
	Mesophyll cell density	0.19		
	Thickness of the upper sclerenchyma layers	0.31		
	Total surface area of bulliform cells	0.07		
	Adaxial stomatal length	0.06		
	Abaxial epidermis thickness	0.06		
	Adaxial epidermis thickness	0.07		
	Leaf mass area	0.15		
	Leaf width	0.21		
	Mean value	0.20	Mean value	0.35

Leaf area (LA), leaf dry mass (DM), leaf mass area (LMA), leaf tissue density (LTD), net photosynthesis (P_N), stomatal conductance (g_s), leaf respiration (R), water use efficiency (WUE), leaf water potential at midday (ψ_m), total chlorophyll content (Chl), relative water content (RWC), gross photosynthesis (P_g), leaf transpiration (E), ratio between leaf respiration, and net photosynthesis (R/P_N) are shown.

COMPETITION BETWEEN INVASIVE AND NATIVE SPECIES

The increase of air temperature and carbon dioxide (CO_2) concentration over recent decades has determined novel environmental conditions [153] which might act as a potent agent of natural selection among species favoring more phenotypically plastic species [154] and resulting in a competition between invasive over cooccurring native species [153, 155]. The phenotypic plasticity of cooccurring native and invasive species in the broadleaf forest developing in the Natural Reserve Siro Negri (45°12′39″N, 9°3′26″E, Italy) attests to the considered species responsiveness to light variations. Q. robur (a native species) and R. pseudoacacia (an invasive species) have a similar physiological plasticity. Nevertheless, the significantly higher morphological and anatomical plasticity of R. pseudoacacia than Q. robur confirms its past capability of colonizing the forest and then growing successfully into both the dominant and dominated layers (data Catoni et al. not published).

Introduced species frequently exhibit little genetic variations in the new environment due to the genetic bottleneck and drift experienced by the small founding populations [156]. Even with genetic variations, local adaptation may not arise if selection is weak or unpredictable, or if considerable gene flow occurs among populations [157]. Despite these limitations, local adaptation often contributes to the success of plants introduced in the new environments [158–162]. In agreement with this hypothesis, Niinemets et al. [42] show higher plasticity of the invasive Rhododendron ponticum in respect to the native Ilex aquifolium in high light environments by its higher N investment in light harvesting and in photosynthetic machinery allowing the production of more leaf area with equal photosynthetic and light-harvesting characteristics. The high phenotypic plasticity in photoprotective strategies and performance of the invasive Taraxacum officinaleenhances its competitive ability in alpine environments [68] considering that light intensity is one of the most changing conditions along altitudinal gradients. The T. officinale plants from higher altitudes, where light conditions are more variable, possess greater plasticity than plants from lower altitudes, suggesting that plasticity in ecophysiological traits will be one of the main strategies to colonize environments with variable light conditions. High adaptability can be due either to the adaptive strategy to cope with resource fluctuations in the native region [163, 164] or to a rapid evolution in novel environments after colonization [165]. Davidson et al. [153] show that invasive species are more plastic in their response to greater resource availability than noninvasives, but this plasticity is only sometimes associated with a fitness benefit, and noninvasive species maintain greater fitness homoeostasis when comparing growth between low and average resource availabilities. Funk [166] investigated the plastic responses of five invasive-native pairs in low resource environments of the Hawaiian Islands and found that the maximum photosynthetic rate and the organic leaf N concentration were positively related to the invasive species fitness in response to N availability. Since adaptive plasticity may allow certain species to colonize environmentally diverse areas without the lag time required for local adaptation, it may enhance their invasiveness and rapid geographic spread contributing to the displacement of native species [4].

Nevertheless, despite the effort over the last decades, the evolutionary mechanisms leading to invasiveness remain unclear [167]. Molina-Montenegro et al. [68] show that invasive species have significantly greater plasticity than native species. On the contrary, Drenovsky et al. [168] suggest that native and invasive species may converge on functionally similar traits demonstrating comparable ability to respond to change in resource availability. In addition, Godoy et al. [169] show that despite reasonable arguments in favour of linking phenotypic plasticity to plant invasion, no general pattern between phenotypic plasticity and invasiveness emerged. DeWalt et al. [170] tested the hypothesis that the tropical shrub Clidemia hirta is more abundant in the introduced (Hawaiian Islands) than in the native range (Costa Rica) because of genetic differences in resource acquisition, allocation, and phenotypic plasticity between native and introduced genotypes. Nevertheless, the results underline the fact that genetic shifts in resource use, resource allocation, or plasticity do not contribute to differences in habitat distribution and abundance. An increased plasticity may not increase fitness (nonadaptive plasticity) or may even decrease it, and the correlation among different plant traits may confers invasiveness by reducing the cost of maladaptive and/or nonadaptive plastic traits [14,171]. Bastlová and Květ [43] evidenced the phenotypic variability in native populations of Lythrum salicaria, a Eurasian species which successfully invaded north American wetlands competing with native plant species [44]. The authors showed that plants originating from more southern Eurasian localities were more similar to the invasive plants in North America than to plants from northern Eurasian localities. Variability in growth characteristics across the north-south gradient within the native range could result from long-term adaptation to prevailing environmental conditions, particularly day-length. Moreover, variability for some growth characteristics (i.e., dry weight and number of lateral branches, root dry weight), both between and within Eurasian populations, indicates a plastic growth response to the local environmental conditions.

CONCLUSIONS

Overall, climate change has been shown to affect abundance and distribution of plant species, as well as plant community composition [172, 173]. Recent studies indicate that under rapid climate change phenotypic plasticity rather than genetic diversity is likely to play a crucial role in allowing plants to persist in their environments [174]. Different responses to climate occur not only between populations throughout a species range but also between cooccurring individuals within a population [149]. Nevertheless, studies of genetic structure based on enzyme polymorphism in populations of forest trees have shown significant levels of intrapopulation variability and little interpopulation differentiation [61, 175, 176]. Increased summer drought will exacerbate the regeneration of many tree species at their lower latitudinal and altitudinal distribution limits and the introduction of more drought tolerant species in vulnerable habitats is considered to facilitate forest persistence [177]. Nevertheless, introducing more drought tolerant species to mitigate climate change might not necessarily be successful due to trade-offs between drought tolerance and growth plasticity. Nicotra et al. [178] sustain the fact that autochthonous provenances have the potential for resistance to change in climatic conditions as a function of both phenotypic plasticity and genotypic variations.

Assessment of vegetation vulnerability and climate change resilience require understanding of the diversity among plant species in the current vegetation and their growth strategies in response to fluctuating resource availability [179]. Since species with extensive geographical range have the potential to exhibit large intraspecific variations in physiology, morphology, and phenology, they may be good models for the study of local and regional adaptations [140]. Nevertheless, adaptation to future global change could require the evolution of a number of different traits that may be constrained by correlations between them [180]. Thus, it is necessary to identify plant functional traits in which plasticity is likely to be a determinant in plant response to global change contributing to predict species distribution changes and shifts [178]. Moreover, it is important to fully understand the ecological consequences at a species and ecosystem level considering that species with a greater adaptive plasticity may be more likely to survive novel environmental conditions, since such changes typically occur too rapidly to allow for an evolutionary (or in some cases a migratory) response. Plasticity is recognized to be a major source of phenotypic variations in the real world because it will influence natural selection and, consequently, patterns of diversification among populations and, ultimately, species [2, 181]. Plasticity promotes evolutionary diversification if the produced phenotypes provide adaptive diversity that under selection becomes evolutionarily fixed [2, 182, 183]. Nevertheless, the extent to which phenotypic plasticity may facilitate survival under changing environmental conditions still remains largely unknown. Although phenotypic plasticity may facilitate short-term adaptation to environmental changes, genetic adaptation might ultimately be necessary for the persistence of species in extreme habitats [177].

In the near future it will be important to collect data by working in the field and, in particular, in primary forests and/or in well-conserved habitats where new stress factors are limited in order to define standard protocols useful for comparative studies. Among the strategies, environment conservation should protect heterogeneity between and within habitats in order to maintain larger intraspecific variability and, thereby conserving a variety of phenotypic specializations that will be able to buffer future environmental extremities due to climate and land-use changes [184].

REFERENCES

1. C. D. Schlichting, "Phenotypic plasticity in plants," Plant Species Biology, vol. 17, no. 2-3, pp. 85–88, 2002.
2. S. E. Sultan, "Promising directions in plant phenotypic plasticity," Perspectives in Plant Ecology, Evolution and Systematics, vol. 6, no. 4, pp. 227–233, 2004.
3. F. Valladares, D. Sanchez-Gomez, and M. A. Zavala, "Quantitative estimation of phenotypic plasticity: bridging the gap between the evolutionary concept and its ecological applications," Journal of Ecology, vol. 94, no. 6, pp. 1103–1116, 2006.
4. W. E. Spencer, J. Teeri, and R. G. Wetzel, "Acclimation of photosynthetic phenotype to environmental heterogeneity," Ecology, vol. 75, no. 2, pp. 301–314, 1994.
5. L. Gratani, P. Pesoli, M. F. Crescente, K. Aichner, and W. Larcher, "Photosynthesis as a temperature indicator in Quercus ilex L," Global and Planetary Change, vol. 24, no. 2, pp. 153–163, 2000. C. Parmesan, "Ecological and evolutionary responses to recent climate change," Annual Review of Ecology, Evolution, and Systematics, vol. 37, pp. 637–669, 2006.
6. S. Matesanz, E. Gianoli, and F. Valladares, "Global change and the evolution of phenotypic plasticity in plants," Annals of the New York Academy of Sciences, vol. 1206, pp. 35–55, 2010.
7. M. van Kleunen and M. Fischer, "Progress in the detection of costs of phenotypic plasticity in plants,"New Phytologist, vol. 176, no. 4, pp. 727–730, 2007.
8. R. Lande, "Adaptation to an extraordinary environment by evolution of phenotypic plasticity and genetic assimilation," Journal of Evolutionary Biology, vol. 22, no. 7, pp. 1435–1446, 2009.
9. A. D. Bradshaw, "Evolutionary significance of phenotypic plasticity in plants," Advances in Genetics, vol. 13, pp. 115–155, 1965.
10. S. M. Scheiner and C. J. Goodnight, "The comparison of phenotypic plasticity and genetic variation in populations of the grass Danthonia spicata," Evolution, vol. 38, pp. 845–855, 1984.

11. S. Via and R. Lande, "Evolution of genetic variability in a spatially heterogeneous environment: effects of genotype-environment interaction," Genetical Research, vol. 49, no. 2, pp. 147–156, 1987.
12. S. Via, R. Gomulkiewicz, G. De Jong, S. M. Scheiner, C. D. Schlichting, and P. H. Van Tienderen, "Adaptive phenotypic plasticity: consensus and controversy," Trends in Ecology and Evolution, vol. 10, no. 5, pp. 212–217, 1995.
13. M. van Kleunen and M. Fischer, "Constraints on the evolution of adaptive phenotypic plasticity in plants," New Phytologist, vol. 166, no. 1, pp. 49–60, 2005.
14. D. D. Ackerly, S. Dudley, S. E. Sultan, et al., "The evolution of plant ecophysiological traits: recent advances and future directions," BioScience, vol. 50, no. 11, pp. 979–995, 2000.
15. T. D. Price, A. Qvarnstrom, and D. E. Irwin, "The role of phenotypic plasticity in driving genetic evolution," Proceedings of the Royal Society B: Biological Sciences, vol. 270, pp. 1433–1440, 2003.
16. L. Gratani, "Leaf and shoot growth dynamics of Quercus ilex L," Acta Oecologica, vol. 17, no. 1, pp. 17–27, 1996.
17. A. Pintado, F. Valladares, and L. G. Sancho, "Exploring phenotypic plasticity in the lichen Ramalina capitata: morphology, water relations and chlorophyll content in North- and South-facing populations,"Annals of Botany, vol. 80, no. 3, pp. 345–353, 1997.
18. L. Gratani, M. Meneghini, P. Pesoli, and M. F. Crescente, "Structural and functional plasticity ofQuercus ilex seedlings of different provenances in Italy," Trees—Structure and Function, vol. 17, no. 6, pp. 515–521, 2003.
19. M. Zunzunegui, M. C. D. Barradas, F. Ain-Lhout, L. Alvarez-Cansino, M. P. Esquivias, and F. G. Novo, "Seasonal physiological plasticity and recovery capacity after summer stress in Mediterranean scrub communities," Plant Ecology, vol. 212, no. 1, pp. 127–142, 2011.
20. D. E. Waitt and D. A. Levin, "Genetic and phenotypic correlations in plants: a botanical test of Cheverud's conjecture," Heredity, vol. 80, no. 3, pp. 310–319, 1998.

21. M. A. Arntz and L. F. Delph, "Pattern and process: evidence for the evolution of photosynthetic traits in natural populations," Oecologia, vol. 127, no. 4, pp. 455–467, 2001.
22. S. E. Sultan, "Phenotypic plasticity for plant development, function and life history," Trends in Plant Science, vol. 5, no. 12, pp. 537–542, 2000.
23. L. Gratani and M. F. Crescente, "Phenology and leaf adaptive strategies of Mediterranean maquis plants," Ecologia Mediterranea, vol. 23, pp. 11–19, 1997.
24. L. A. Dorn, E. H. Pyle, and J. Schmitt, "Plasticity to light cues and resources in Arabidopsis thaliana: testing for adaptive value and costs," Evolution, vol. 54, no. 6, pp. 1982–1994, 2000.
25. F. Valladares, E. Gianoli, and J. M. Gómez, "Ecological limits to plant phenotypic plasticity," New Phytologist, vol. 176, no. 4, pp. 749–763, 2007.
26. T. Wyka, P. Robakowski, and R. Zytkowiak, "Acclimation of leaves to contrasting irradiance in juvenile trees differing in shade tolerance," Tree Physiology, vol. 27, no. 9, pp. 1293–1306, 2007.
27. M. B. Walters and P. B. Reich, "Seed size, nitrogen supply, and growth rate affect tree seedling survival in deep shade," Ecology, vol. 81, no. 7, pp. 1887–1901, 2000.
28. J. Stöcklin, P. Kuss, and A. R. Pluess, "Genetic diversity, phenotypic variation and local adaptation in the alpine landscape: case studies with alpine plant species," Botanica Helvetica, vol. 119, no. 2, pp. 125–133, 2009.
29. F. S. Matos, R. Wolfgramm, F. V. Gonçalves, P. C. Cavatte, M. C. Ventrella, and F. M. DaMatta, "Phenotypic plasticity in response to light in the coffee tree," Environmental and Experimental Botany, vol. 67, no. 2, pp. 421–427, 2009.
30. W. L. Araujo, P. C. Dias, G. A. B. K. Moraes et al., "Limitations to photosynthesis in coffee leaves from different canopy positions," Plant Physiology and Biochemistry, vol. 46, no. 10, pp. 884–890, 2008.
31. Ü. Niinemets, O. Kull, and J. D. Tenhunen, "An analysis of light effects on foliar morphology, physiology, and light interception in temperate deciduous woody species of contrasting shade tolerance," Tree Physiology, vol. 18, no. 10, pp. 681–696, 1998.

32. H. Ellenberg, Vegetation Ecology of Central Europe, Cambridge University Press, Cambridge, UK, 4th edition, 1988.

33. H. J. Otto, Waldökologie, Eugen Ulmer, Stuttgart, Germany, 1994.

34. L. Gratani, R. Catoni, G. Pirone, A. R. Frattaroli, and L. Varone, "Physiological and morphological leaf trait variations in two Apennine plant species in response to different altitudes," Photosynthetica, vol. 50, no. 1, pp. 15–23, 2012.

35. G. Ke and M. J. A. Werger, "Different responses to shade of evergreen and deciduous oak seedlings and the effect of acorn size," Acta Oecologica, vol. 20, no. 6, pp. 579–586, 1999.

36. F. Valladares, J. Chico, I. Aranda et al., "The greater seedling high-light tolerance of Quercus robur overFagus sylvatica is linked to a greater physiological plasticity," Trees—Structure and Function, vol. 16, no. 6, pp. 395–403, 2002.

37. G. Müller-Stark, Biodervisität und Nachhaltige Forstwirtschaft, Ecomed, Landsberg, Germany, 1997.

38. A. D. Peuke, C. Schraml, W. Hartung, and H. Rennenberg, "Identification of drought-sensitive beech ecotypes by physiological parameters," New Phytologist, vol. 154, no. 2, pp. 373–387, 2002.

39. Z. Wen, Y. Ding, T. Zhao, and J. Gai, "Genetic diversity and peculiarity of annual wild soybean (G. sojaSieb. et Zucc.) from various eco-regions in China," Theoretical and Applied Genetics, vol. 119, no. 2, pp. 371–381, 2009.

40. L. Sack, P. J. Grubb, and T. Marañón, "The functional morphology of juvenile plants tolerant of strong summer drought in shaded forest understories in southern Spain," Plant Ecology, vol. 168, no. 1, pp. 139–163, 2003.

41. Ü. Niinemets, F. Valladares, and R. Ceulemans, "Leaf-level phenotypic variability and plasticity of invasive Rhododendron ponticum and non-invasive Ilex aquifolium co-occurring at two contrasting European sites," Plant, Cell and Environment, vol. 26, no. 6, pp. 941–956, 2003.

42. D. Bastlová and J. Květ, "Phenotypic plasticity in native populations of Lythrum salicaria L. across geographical gradient: between- and within-population differences," in Ecology and

Management of Alien Plant Invasions, L. Child, Ed., pp. 237–247, Backhuys, Leiden, The Netherlands, 2003.

43. D. Q. Thompson, R. L. Stuckey, and E. B. Thompson, "Spread, impact, and control of purple loosestrife (Lythrum salicaria) in North American wetlands," U.S. Fish and Wildlife Service, Fish and Wildlife Research Report 2, Washington, DC, USA, 1987.
44. M. M. Mendes, L. C. Gazarini, and M. L. Rodrigues, "Acclimation of Myrtus communis to contrasting Mediterranean light environments—effects on structure and chemical composition of foliage and plant water relations," Environmental and Experimental Botany, vol. 45, no. 2, pp. 165–178, 2001.
45. L. Gratani, F. Covone, and W. Larcher, "Leaf plasticity in response to light of three evergreen species of the Mediterranean maquis," Trees—Structure and Function, vol. 20, no. 5, pp. 549–558, 2006.
46. J. Oleksyn, J. Modrzyński, M. G. Tjoelker, R. Zytkowiak, P. B. Reich, and P. Karolewski, "Growth and physiology of Picea abies populations from elevational transects: common garden evidence for altitudinal ecotypes and cold adaptation," Functional Ecology, vol. 12, no. 4, pp. 573–590, 1998.
47. L. Poorter, E. Lianes, M. Moreno-de las Heras, and M. A. Zavala, "Architecture of Iberian canopy tree species in relation to wood density, shade tolerance and climate," Plant Ecology, vol. 213, no. 5, pp. 707–722, 2012.
48. M. A. Zavala, J. M. Espelta, and J. Retana, "Constraints and trade-offs in Mediterranean plant communities: the case of holm oak-Aleppo pine forests," Botanical Review, vol. 66, no. 1, pp. 119–149, 2000.
49. F. J. Baquedano, F. Valladares, and F. J. Castillo, "Phenotypic plasticity blurs ecotypic divergence in the response of Quercus coccifera and Pinus halepensis to water stress," European Journal of Forest Research, vol. 127, no. 6, pp. 495–506, 2008.
50. S. Palmroth, F. Berninger, E. Nikinmaa, J. Lloyd, P. Pulkkinen, and P. Hari, "Structural adaptation rather than water conservation was observed in Scots pine over a range of wet to dry climates,"Oecologia, vol. 121, no. 3, pp. 302–309, 1999.

51. S. Nahum, M. Inbar, G. Ne'eman, and R. Ben-Shlomo, "Phenotypic plasticity and gene diversity inPistacia lentiscus L. along environmental gradients in Israel," Tree Genetics and Genomes, vol. 4, no. 4, pp. 777–785, 2008.
52. S. E. Sultan, "An emerging focus on plant ecological development," New Phytologist, vol. 166, no. 1, pp. 1–5, 2005.
53. L. Gratani, R. Catoni, and L. Varone, "Morphological, anatomical and physiological leaf traits of Q. ilex,P. latifolia, P. lentiscus, and M. communis and their response to Mediterranean climate stress factors,"Botanical Studies, vol. 54, pp. 1–12, 2013.
54. L. Gratani, E. Fiorentino, A. Kubova, and P. Marzi, "Effect of microclimate on ecophysiological features of some sclerophyllous species," Photosynthetica, vol. 23, pp. 230–233, 1989.
55. L. Balaguer, E. Martínez-Ferri, F. Valladares et al., "Population divergence in the plasticity of the response of Quercus coccifera to the light environment," Functional Ecology, vol. 15, no. 1, pp. 124–135, 2001.
56. A. Bonito, L. Varone, and L. Gratani, "Relationship between acorn size and seedling morphological and physiological traits of Quercus ilex L. from different climates," Photosynthetica, vol. 49, no. 1, pp. 75–86, 2011.
57. P. Pesoli, L. Gratani, and W. Larcher, "Responses of Quercus ilex from different provenances to experimentally imposed water stress," Biologia Plantarum, vol. 46, no. 4, pp. 577–581, 2003.
58. M. Barbero, R. Loisel, and P. Quézel, "Biogeography, ecology and history of Mediterranean Quercus ilex ecosystems," Vegetatio, vol. 99-100, no. 1, pp. 19–34, 1992.
59. J. Terradas and R. Savé, "The influence of summer and winter stress and water relationship on the distribution of Quercus ilex L.," in Quercus ilex L. Ecosystems: Function, Dynamics and Management, F. Romane and J. Terradas, Eds., pp. 137–145, Kluwer Academic, Dodrecht, The Netherlands, 1992.
60. H. Michaud, R. Lumaret, and F. Romane, "Variation in the genetic structure and reproductive biology of holm oak populations," Vegetatio, vol. 99-100, no. 1, pp. 107–113, 1992.
61. K. Hayashi, "Growth characteristics and silviculture in Quercus mongolica var. crispula Blume," inHandbook for Hardwood, K.

Sakaguchi, Ed., pp. 122–128, Association For Promotion of Forest Science, Tokyo, Japan, 1985.

62. M. Higo, "Regeneration behaviors of tree species of secondary stands regenerated on sites disturbed by Typhoon 15: based on the proportion of advanced regeneration, growth rate, and seedling density in closed mature stands," Journal of the Japanese Forestry Society, vol. 76, pp. 531–539, 1994.
63. M.-L. Navas and E. Garnier, "Plasticity of whole plant and leaf traits in Rubia peregrina in response to light, nutrient and water availability," Acta Oecologica, vol. 23, no. 6, pp. 375–383, 2002.
64. L. Gratani, M. F. Crescente, V. D'amato, C. Ricotta, A. R. Frattaroli, and G. Puglielli, "Leaf traits variation in Sesleria nitida growing at different altitudes in the Central Apennines," Photosynthetica. In press.
65. P. M. S. Ashton and G. P. Berlyn, "Leaf adaptations of some Shorea species to sun and shade," New Phytologist, vol. 121, no. 4, pp. 587–596, 1992.
66. R. J. N. Emery, D. M. Reid, and C. C. Chinnappa, "Phenotypic plasticity of stem elongation in two ecotypes of Stellaria longipes: the role of ethylene and response to wind," Plant, Cell and Environment, vol. 17, no. 6, pp. 691–700, 1994.
67. M. A. Molina-Montenegro, J. Peñuelas, S. Munné-Bosch, and J. Sardans, "Higher plasticity in ecophysiological traits enhances the performance and invasion success of Taraxacum officinale(dandelion) in alpine environments," Biological Invasions, vol. 14, no. 1, pp. 21–33, 2012.
68. W. Larcher, Physiological Plant Ecology, Springer, Berlin, Germany, 4th edition, 2003.
69. P. Doughty, "Testing the ecological correlates of phenotypically plastic traits within a phylogenetic framework," Acta Oecologica, vol. 16, no. 4, pp. 519–524, 1995.
70. K. Gotthard and S. Nylin, "Adaptive plasticity and plasticity as an adaptation: a selective review of plasticity in animal morphology and life history," Oikos, vol. 74, no. 1, pp. 3–17, 1995.
71. S. A. Cook and M. P. Johnson, "Adaptation to heterogeneous environments. I. Variation in heterophylly in Ranunculus flammula," Evolution, vol. 22, pp. 496–516, 1968.

72. T. O. Haugen and L. A. Vollestad, "Population differences in early life-history traits in grayling,"Journal of Evolutionary Biology, vol. 13, no. 6, pp. 897–905, 2000.

73. C. K. Ghalambor and T. E. Martin, "Comparative manipulation of predation risk in incubating birds reveals variability in the plasticity of responses," Behavioral Ecology, vol. 13, no. 1, pp. 101–108, 2002.

74. M. Pigliucci, C. J. Murren, and C. D. Schlichting, "Phenotypic plasticity and evolution by genetic assimilation," Journal of Experimental Biology, vol. 209, no. 12, pp. 2362–2367, 2006.

75. A. Weigelt and P. Jolliffe, "Indices of plant competition," Journal of Ecology, vol. 91, no. 5, pp. 707–720, 2003.

76. P. E. Hulme, "Phenotypic plasticity and plant invasions: is it all Jack?" Functional Ecology, vol. 22, no. 1, pp. 3–7, 2008.

77. D. Sánchez-Gómez, F. Valladares, and M. A. Zavala, "Functional traits and plasticity in response to light in seedlings of four Iberian forest tree species," Tree Physiology, vol. 26, no. 11, pp. 1425–1433, 2006.

78. H. Poorter, Ü. Niinemets, A. Walter, F. Fiorani, and U. Schurr, "A method to construct dose-response curves for a wide range of environmental factors and plant traits by means of a meta-analysis of phenotypic data," Journal of Experimental Botany, vol. 61, no. 8, pp. 2043–2055, 2010.

79. D. Kuiper and P. J. C. Kuiper, "Phenotypic plasticity in a physiological perspective," Acta Oecologica Oecologia Plantarum, vol. 9, pp. 43–59, 1988.

80. F. Valladares, L. Balaguer, E. Martinez-Ferri, E. Perez-Corona, and E. Manrique, "Plasticity, instability and canalization: Is the phenotypic variation in seedlings of sclerophyll oaks consistent with the environmental unpredictability of Mediterranean ecosystems?" New Phytologist, vol. 156, no. 3, pp. 457–467, 2002.

81. Ü. Niinemets and F. Valladares, "Photosynthetic acclimation to simultaneous and interacting environmental stresses along natural light gradients: optimality and constraints," Plant Biology, vol. 6, no. 3, pp. 254–268, 2004.

82. M. Zunzunegui, F. Ain-Lhout, M. C. D. Barradas, L. Álvarez-Cansino, M. P. Esquivias, and F. García Novo, "Physiological, morphological and allocation plasticity of a semi-deciduous shrub," Acta Oecologica, vol. 35, no. 3, pp. 370–379, 2009.
83. F. S. Chapin III, K. Autumn, and F. Pugnaire, "Evolution of suites of traits in response to environmental stress," American Naturalist, vol. 142, pp. S78–S92, 1993.
84. F. Valladares, E. Martinez-Ferri, L. Balaguer, E. Perez-Corona, and E. Manrique, "Low leaf-level response to light and nutrients in Mediterranean evergreen oaks: a conservative resource-use strategy?"New Phytologist, vol. 148, no. 1, pp. 79–91, 2000.
85. J. C. Crick and J. P. Grime, "Morphological plasticity and mineral nutrient capture in two herbaceous species of contrasted ecology," New Phytologist, vol. 107, no. 2, pp. 403–414, 1987.
86. N. G. Dengler, "Comparative histological basis of sun and shade leaf dimorphism in Helianthus annuus," Canadian Journal of Botany, vol. 58, pp. 717–730, 1980.
87. J. H. McClendon and G. G. McMillen, "The control of leaf morphology and the tolerance of shade by woody plants," Botanical Gazette, vol. 143, no. 1, pp. 79–83, 1982.
88. M. P. Dale and D. R. Causton, "The ecophysiology of Veronica chamaedrys, V. montana and V. officinalis. I. Light quality and light quantity," Journal of Ecology, vol. 80, no. 3, pp. 483–492, 1992.
89. J. Květ, J. Svoboda, and K. Fiala, "Canopy development in stands of Typha latifolia L. and Phragmites communis Trin. in South Moravia," Hidrobiologia, vol. 10, pp. 63–75, 1969.
90. J. R. Evans, "Leaf anatomy enables more equal access to light and CO_2 between chloroplasts," New Phytologist, vol. 143, no. 1, pp. 93–104, 1999.
91. R. B. Wylie, "Principles of foliar organization shown by sun-shade leaves from ten species of deciduous dicotyledonous trees," American Journal of Botany, vol. 36, pp. 355–361, 1951.
92. J. H. McClendon, "The relationship between the thickness of deciduous leaves and their maximum photosynthetic rate," American Journal of Botany, vol. 49, pp. 320–322, 1962.

93. H. K. Lichtenthaler, C. Buschmann, M. Döll et al., "Photosynthetic activity, chloroplast ultrastructure, and leaf characteristics of high-light and low-light plants and of sun and shade leaves," Photosynthesis Research, vol. 2, no. 2, pp. 115–141, 1981.
94. H. K. Lichtenthaler, G. Kuhn, U. Prenzel, and D. Meier, "Chlorophyll-protein levels and degree of thylakoid stacking in radish chloroplasts from high-light, low-light and bentazon-treated plants,"Physiologia Plantarum, vol. 56, no. 2, pp. 183–188, 1982.
95. J. R. Evans, "Photosynthesis and nitrogen relationships in leaves of C_3 plants," Oecologia, vol. 78, no. 1, pp. 9–19, 1989.
96. K. Hikosaka, "Nitrogen partitioning in the photosynthetic apparatus of Plantago asiatica leaves grown under different temperature and light conditions: similarities and differences between temperature and light acclimation," Plant and Cell Physiology, vol. 46, no. 8, pp. 1283–1290, 2005.
97. T. Hirose and M. J. A. Werger, "Maximizing daily canopy photosynthesis with respect to the leaf nitrogen allocation pattern in the canopy," Oecologia, vol. 72, no. 4, pp. 520–526, 1987.
98. K. Yoshimura, "Irradiance heterogeneity within crown affects photosynthetic capacity and nitrogen distribution of leaves in Cedrela sinensis," Plant, Cell and Environment, vol. 33, no. 5, pp. 750–758, 2010.
99. N. P. R. Anten, "Modelling canopy photosynthesis using parameters determined from simple non-destructive measurements," Ecological Research, vol. 12, no. 1, pp. 77–88, 1997.
100. L. Sack, P. J. Melcher, W. H. Liu, E. Middleton, and T. Pardee, "How strong is intracanopy leaf plasticity in temperate deciduous trees?" American Journal of Botany, vol. 93, no. 6, pp. 829–839, 2006.
101. Ü. Niinemets, "A review of light interception in plant stands from leaf to canopy in different plant functional types and in species with varying shade tolerance," Ecological Research, vol. 25, no. 4, pp. 693–714, 2010.
102. H. P. Meister, M. M. Caldwell, J. D. Tenhunen, and O. L. Lange, "Ecological implications of sun/shade-leaf differentiation in sclerophyllous canopies: assessment by canopy modeling," in

Plant Response to Stress: Functional Analysis in Mediterranean Ecosystems, J. D. Tenhunen, F. M. Catarino, O. L. Lange, and W. C. Oechel, Eds., pp. 401–411, Springer, Heidelberg, Germany, 1987.

103. Ü. Niinemets and O. Kull, "Stoichiometry of foliar carbon constituents varies along light gradients in temperate woody canopies: implications for foliage morphological plasticity," Tree Physiology, vol. 18, no. 7, pp. 467–479, 1998.
104. F. Valladares, S. J. Wright, E. Lasso, K. Kitajima, and R. W. Pearcy, "Plastic phenotypic response to light of 16 congeneric shrubs from a panamanian rainforest," Ecology, vol. 81, no. 7, pp. 1925–1936, 2000.
105. T. J. Givnish, "Adaptive significance of evergreen vs. deciduous leaves: solving the triple paradox,"Silva Fennica, vol. 36, no. 3, pp. 703–743, 2002.
106. L. Markesteijn, L. Poorter, and F. Bongers, "Light-dependent leaf trait variation in 43 tropical dry forest tree species," American Journal of Botany, vol. 94, no. 4, pp. 515–525, 2007.
107. T. P. Wyka, J. Oleksyn, R. Zytkowiak, P. Karolewski, A. M. Jagodziński, and P. B. Reich, "Responses of leaf structure and photosynthetic properties to intra-canopy light gradients: a common garden test with four broadleaf deciduous angiosperm and seven evergreen conifer tree species," Oecologia, vol. 170, pp. 11–24, 2012.
108. L. Gratani and I. Foti, "Estimating forest structure and shade tolerance of the species in a mixed deciduous broad-leaved forest in Abruzzo, Italy," Annales Botanici Fennici, vol. 35, no. 2, pp. 75–83, 1998.
109. L. Gratani, "Canopy structure, vertical radiation profile and photosynthetic function in a Quercus ilexevergreen forest," Photosynthetica, vol. 33, no. 1, pp. 139–149, 1997.
110. R. K. P. Yadav, A. M. Bosabalidis, and D. Vokou, "Leaf structural features of Mediterranean perennial species: plasticity and life form specificity," Journal of Biological Research, vol. 2, pp. 21–34, 2004.
111. F. Valladares and Ü. Niinemets, "Feature of complex nature and consequences," Annual Review of Ecology, Evolution, and Systematics, vol. 39, pp. 237–257, 2008.

112. S. B. Carpenter and N. D. Smith, "A comparative studies of leaf thickness among Southern Appalachian hardwoods," Canadian Journal of Botany, vol. 59, pp. 1393–1396, 1981.
113. S. Strauss-Debenedetti and F. A. Bazzaz, "Plasticity and acclimation to light in tropical Moraceae of different sucessional positions," Oecologia, vol. 87, no. 3, pp. 377–387, 1991.
114. C. C. Muth and F. A. Bazzaz, "Tree canopy displacement at forest gap edges," Canadian Journal of Forest Research, vol. 32, no. 2, pp. 247–254, 2002.
115. F. Longuetaud, A. Piboule, H. Wernsdörfer, and C. Collet, "Crown plasticity reduces inter-tree competition in a mixed broadleaved forest," European Journal of Forest Research, vol. 132, pp. 621–634, 2013.
116. M. H. Turnbull, "The effect of light quantity and quality during development on the photosynthetic characteristics of six Australian rainforest tree species," Oecologia, vol. 87, no. 1, pp. 110–117, 1991.
117. J. Popma, F. Bongers, and M. J. A. Werger, "Gap-dependence and leaf characteristics of trees in a tropical lowland rain forest in Mexico," Oikos, vol. 63, no. 2, pp. 207–214, 1992.
118. F. Valladares, S. Arrieta, I. Aranda et al., "Shade tolerance, photoinhibition sensitivity and phenotypic plasticity of Ilex aquifolium in continental Mediterranean sites," Tree Physiology, vol. 25, no. 8, pp. 1041–1052, 2005.
119. R. Catoni, L. Gratani, and L. Varone, "Physiological, morphological and anatomical trait variations between winter and summer leaves of Cistus species," Flora: Morphology, Distribution, Functional Ecology of Plants, vol. 207, pp. 442–449, 2012.
120. A. F. M. van Hees, "Growth and morphology of pedunculate oak (Quercus robur L) and beech (Fagus sylvatica L) seedlings in relation to shading and drought," Annales des Sciences Forestieres, vol. 54, no. 1, pp. 9–18, 1997.
121. H. Ishii and S. Asano, "The role of crown architecture, leaf phenology and photosynthetic activity in promoting complementary use of light among coexisting species in temperate forests," Ecological Research, vol. 25, no. 4, pp. 715–722, 2010.

122. A. M. Barton, "Factors controlling plant distributions: drought, competition, and fire in montane pines in Arizona," Ecological Monographs, vol. 63, no. 4, pp. 367–397, 1993.
123. J. Martínez-Vilalta and J. Piñol, "Drought-induced mortality and hydraulic architecture in pine populations of the NE Iberian Peninsula," Forest Ecology and Management, vol. 161, no. 1–3, pp. 247–256, 2002.
124. M. A. Zavala, J. M. Espelta, J. Caspersen, and J. Retana, "Interspecific differences in sapling performance with respect to light and aridity gradients in mediterranean pine-oak forests: implications for species coexistence," Canadian Journal of Forest Research, vol. 41, no. 7, pp. 1432–1444, 2011.
125. M. Kitao, T. T. Lei, T. Koike, H. Tobita, and Y. Maruyama, "Susceptibility to photoinhibition of three deciduous broadleaf tree species with different successional traits raised under various light regimes,"Plant, Cell and Environment, vol. 23, no. 1, pp. 81–89, 2000.
126. S. S. Mulkey and R. W. Pearcy, "Interactions between acclimation and photoinhibition of photosynthesis of a tropical forest understory herb, Alocasia macrorrhiza, during simulated canopy gap formation," Functional Ecology, vol. 6, pp. 719–729, 1992.
127. C. B. Osmond, "What is photoinhibition? Some insights from comparison of shade and sun plants," inPhotoinhibition of Photosynthesis: from Molecular Mechanisms to the Field, N. R. Baker and J. R. Bowyer, Eds., pp. 1–24, BIOS Scientific Publishers, Lancaster, UK, 1994.
128. T. Koike, M. Kitao, Y. Maruyama, S. Mori, and T. T. Lei, "Leaf morphology and photosynthetic adjustments among deciduous broad-leaved trees within the vertical canopy profile," Tree Physiology, vol. 21, no. 12-13, pp. 951–958, 2001.
129. T. Koike, "Photosynthetic responses to light intensity of deciduous broad-leaved tree seedlings raised under various artificial shade," Environmental Control in Biology, vol. 24, pp. 51–58, 1986.
130. N. C. Vance and J. B. Zaerr, "Influence of drought stress and low irradiance on plant water relations and structural constituents in needles of Pinus ponderosa seedlings," Tree Physiology, vol. 8, pp. 175–184, 1991.

131. S. H. Janse-Ten Klooster, E. J. P. Thomas, and F. J. Sterck, "Explaining interspecific differences in sapling growth and shade tolerance in temperate forests," Journal of Ecology, vol. 95, no. 6, pp. 1250–1260, 2007.
132. T. J. Kawecki and D. Ebert, "Conceptual issues in local adaptation," Ecology Letters, vol. 7, no. 12, pp. 1225–1241, 2004.
133. O. Savolainen, T. Pyhäjärvi, and T. Knürr, "Gene flow and local adaptation in forest trees," Annual Review of Ecology, Evolution, and Systematics, vol. 37, pp. 595–619, 2007.
134. K. M. Hufford and S. J. Mazer, "Plant ecotypes: genetic differentiation in the age of ecological restoration," Trends in Ecology and Evolution, vol. 18, no. 3, pp. 147–155, 2003.
135. M. C. Grant and J. B. Mitton, "Genetic differentiation among growth forms of Engelmann spruce and subalpine fir at tree line," Arctic Antarctic and Alpine Research, vol. 9, pp. 259–263, 1977.
136. J. B. Mitton, Y. B. Linhart, J. L. Hamrick, and J. S. Beckman, "Observations on the genetic structure and mating system of ponderosa pine in the Colorado front range," Theoretical and Applied Genetics, vol. 51, no. 1, pp. 5–13, 1977.
137. C. K. Kelly, M. W. Chase, A. de Bruijn, M. F. Fay, and F. I. Woodward, "Temperature-based population segregation in birch," Ecology Letters, vol. 6, no. 2, pp. 87–89, 2003.
138. B. C. Bongarten and R. O. Teskey, "Water relations of loblolly pine seedlings from diverse geographic origins," Tree Physiology, vol. 1, pp. 265–276, 1986.
139. R. Y. Soolanayakanahally, R. D. Guy, S. N. Silim, E. C. Drewes, and W. R. Schroeder, "Enhanced assimilation rate and water use efficiency with latitude through increased photosynthetic capacity and internal conductance in balsam poplar (Populus balsamifera L.)," Plant, Cell and Environment, vol. 32, no. 12, pp. 1821–1832, 2009.
140. J. A. Ramírez-Valiente, F. Valladares, L. Gil, and I. Aranda, "Population differences in juvenile survival under increasing drought are mediated by seed size in cork oak (Quercus suber L.)," Forest Ecology and Management, vol. 257, no. 8, pp. 1676–1683, 2009.

141. L. Gratani and M. Amadori, "Post-fire resprouting of shrubby species in Mediterranean maquis,"Vegetatio, vol. 96, no. 2, pp. 137–143, 1991.
142. W. Larcher, C. Kainmüller, and J. Wagner, "Survival types of high mountain plants under extreme temperatures," Flora: Morphology, Distribution, Functional Ecology of Plants, vol. 205, no. 1, pp. 3–18, 2010.
143. D. Nogués-Bravo, M. B. Araújo, M. P. Errea, and J. P. Martínez-Rica, "Exposure of global mountain systems to climate warming during the 21st Century," Global Environmental Change, vol. 17, no. 3-4, pp. 420–428, 2007.
144. W. Thuiller, S. Lavorel, M. B. Araújo, M. T. Sykes, and I. C. Prentice, "Climate change threats to plant diversity in Europe," Proceedings of the National Academy of Sciences of the United States of America, vol. 102, no. 23, pp. 8245–8250, 2005.
145. R. Engler, C. F. Randin, W. Thuiller et al., "21st century climate change threatens mountain flora unequally across Europe," Global Change Biology, vol. 17, no. 7, pp. 2330–2341, 2011.
146. T. Dirnböck, F. Essl, and W. Rabitsch, "Disproportional risk for habitat loss of high-altitude endemic species under climate change," Global Change Biology, vol. 17, no. 2, pp. 990–996, 2011.
147. O. Savolainen, F. Bokma, R. García-Gil, P. Komulainen, and T. Repo, "Genetic variation in cessation of growth and frost hardiness and consequences for adaptation of Pinus sylvestris to climatic changes,"Forest Ecology and Management, vol. 197, no. 1–3, pp. 79–89, 2004.
148. A. S. Jump and J. Peñuelas, "Running to stand still: adaptation and the response of plants to rapid climate change," Ecology Letters, vol. 8, no. 9, pp. 1010–1020, 2005.
149. H. Pauli, M. Gottfried, and G. Grabherr, "Effect of climate change on the alpine and nival vegetation of the Alps," Journal of Mountain Ecology, vol. 7, pp. 9–12, 2003.
150. M. Lindner, M. Maroschek, S. Netherer et al., "Climate change impacts, adaptive capacity, and vulnerability of European forest ecosystems," Forest Ecology and Management, vol. 259, no. 4, pp. 698–709, 2010.

151. J. F. Scheepens, E. S. Frei, and J. Stöcklin, "Genotypic and environmental variation in specific leaf area in a widespread Alpine plant after transplantation to different altitudes," Oecologia, vol. 164, no. 1, pp. 141–150, 2010.
152. A. M. Davidson, M. Jennions, and A. B. Nicotra, "Do invasive species show higher phenotypic plasticity than native species and, if so, is it adaptive? A meta-analysis," Ecology Letters, vol. 14, no. 4, pp. 419–431, 2011.
153. S. L. Chown, S. Slabber, M. A. McGeoch, C. Janion, and H. P. Leinaas, "Phenotypic plasticity mediates climate change responses among invasive and indigenous arthropods," Proceedings of the Royal Society B: Biological Sciences, vol. 274, no. 1625, pp. 2531–2537, 2007.
154. J. S. Dukes, "Tomorrow's plant communities: different, but how?" New Phytologist, vol. 176, no. 2, pp. 235–237, 2007.
155. S. C. H. Barrett and B. J. Richardson, "Genetic attributes of invading species," in Ecology of Biological Invasions, R. H. Groves and J. J. Burdon, Eds., pp. 21–33, 1986.
156. J. Antonovics, "The nature of limits to natural selection," Annals of the Missouri Botanical Garden, vol. 63, pp. 224–247, 1976.
157. S. C. H. Barrett, "Genetic variation in weeds," in Biological Control of Weeds with Plant Pathogens, R. Charudattan and H. Walker, Eds., pp. 73–98, John Wiley & Sons, New York, NY, USA, 1982.
158. K. J. Rice and R. N. Mack, "Ecological genetics of Bromus tectorum—II. Intraspecific variation in phenotypic plasticity," Oecologia, vol. 88, no. 1, pp. 84–90, 1991.
159. R. J. Abbott, "Plant invasions, interspecific hybridization and the evolution of new plant taxa," Trends in Ecology and Evolution, vol. 7, no. 12, pp. 401–405, 1992.
160. D. G. Williams and R. A. Black, "Phenotypic variation in contrasting temperature environments: growth and photosynthesis in Pennisetum setaceum from different altitudes on Hawaii," Functional Ecology, vol. 7, no. 5, pp. 623–633, 1993.
161. L. Z. Durand and G. Goldstein, "Photo-synthesis, photoinhibition, and nitrogen use efficiency in native and invasive tree ferns in Hawaii," Oecologia, vol. 126, no. 3, pp. 345–354, 2001.

162. D. G. Williams, R. N. Mack, and R. A. Black, "Ecophysiology of introduced Pennisetum setaceum on Hawaii: the role of phenotypic plasticity," Ecology, vol. 76, no. 5, pp. 1569–1580, 1995.

163. O. K. Atkin, B. R. Loveys, L. J. Atkinson, and T. L. Pons, "Phenotypic plasticity and growth temperature: understanding interspecific variability," in Meeting on Phenotypic Plasticity and the Changing Environment Held at the Society for Experimental Biology Plant Frontiers Meeting, pp. 267–281, Sheffield, UK, 2005.

164. P. J. Yeh and T. D. Price, "Adaptive phenotypic plasticity and the successful colonization of a novel environment," American Naturalist, vol. 164, no. 4, pp. 531–542, 2004.

165. J. L. Funk, "Differences in plasticity between invasive and native plants from a low resource environment," Journal of Ecology, vol. 96, no. 6, pp. 1162–1173, 2008.

166. S. Lavergne and J. Molofsky, "Increased genetic variation and evolutionary potential drive the success of an invasive grass," Proceedings of the National Academy of Sciences of the United States of America, vol. 104, no. 10, pp. 3883–3888, 2007.

167. R. E. Drenovsky, B. J. Grewell, C. M. D'Antonio et al., "A functional trait perspective on plant invasions," Annals of Botany, vol. 110, pp. 141–153, 2012.

168. O. Godoy, A. Saldana, N. Fuentes, F. Valladares, and E. Gianoli, "Forests are not immune to plant invasions: phenotypic plasticity and local adaptation allow Prunella vulgaris to colonize a temperate evergreen rainforest," Biological Invasions, vol. 13, no. 7, pp. 1615–1625, 2011.

169. S. J. DeWalt, J. S. Denslow, and J. L. Hamrick, "Biomass allocation, growth, and photosynthesis of genotypes from native and introduced ranges of the tropical shrub Clidemia hirta," Oecologia, vol. 138, no. 4, pp. 521–531, 2004.

170. P. Poot and H. Lambers, "Shallow-soil endemics: adaptive advantages and constraints of a specialized root-system morphology," New Phytologist, vol. 178, no. 2, pp. 371–381, 2008.

171. G.-R. Walther, E. Post, P. Convey et al., "Ecological responses to recent climate change," Nature, vol. 416, no. 6879, pp. 389–395, 2002.

172. A. Menzel, T. H. Sparks, N. Estrella, et al., "European phenological response to climate change matches the warming pattern," Global Change Biology, vol. 12, pp. 1–8, 2006.

173. Y. Vitasse, C. C. Bresson, A. Kremer, R. Michalet, and S. Delzon, "Quantifying phenological plasticity to temperature in two temperate tree species," Functional Ecology, vol. 24, no. 6, pp. 1211–1218, 2010.

174. M. Desponts and J. P. Simon, "Structure et variabilité génétique de populations d'épinettes noires(Picea mariana (Mill.) B.S.P.) dans la zone hémiarctique du Nouveau- Quèbec," Canadian Journal of Forest Research, vol. 17, pp. 1006–1012, 1987.

175. F. Villani, S. Benedettelli, M. Paciucci, M. Cherubini, and M. Pigliucci, "Genetic variation and differentiation between natural populations or chestnust (Castanea sativa Mill.) from Italy," inBiochemical Markers in the Population Genetics of Forest Trees, H. H. Hattemer, S. Fineschi, F. Cannata, and M. E. Malvolti, Eds., pp. 91–103, SPB Acadamic, The Hague, The Netherlands, 1991.

176. S. Richter, T. Kipfer, T. Wohlgemuth, C. C. Guerrero, J. Ghazoul, and B. Moser, "Phenotypic plasticity facilitates resistance to climate change in a highly variable environment," Oecologia, vol. 169, no. 1, pp. 269–279, 2012.

177. A. B. Nicotra, O. K. Atkin, S. P. Bonser et al., "Plant phenotypic plasticity in a changing climate,"Trends in Plant Science, vol. 15, no. 12, pp. 684–692, 2010.

178. I. K. Dawson, A. Lengkeek, J. C. Weber, and R. Jamnadass, "Managing genetic variation in tropical trees: Linking knowledge with action in agroforestry ecosystems for improved conservation and enhanced livelihoods," Biodiversity and Conservation, vol. 18, no. 4, pp. 969–986, 2009.

179. J. R. Etterson and R. G. Shaw, "Constraint to adaptive evolution in response to global warming,"Science, vol. 294, no. 5540, pp. 151–154, 2001.

180. S. M. Scheiner, "The genetics of phenotypic plasticity. VII. Evolution in a spatially-structured environment," Journal of Evolutionary Biology, vol. 11, no. 3, pp. 303–320, 1998.

181. G. G. Simpson, "The Baldwin effect," Evolution, vol. 7, pp. 110–117, 1953.

182. M. Pigliucci and C. J. Murren, "Genetic assimilation and a possible evolutionary paradox: can macroevolution sometimes be so fast as to pass us by?" Evolution, vol. 57, no. 7, pp. 1455–1464, 2003.

183. C. Wellstein, S. Chelli, G. Campetella et al., "Intraspecific phenotypic variability of plant functional traits in contrasting mountain grassland habitats," Biodiversity and Conservation, vol. 22, pp. 2353–2374, 2013.

Chapter 6

Allopolyploidy and the Evolution of Plant Virus Resistance

John Gottula[1], Ramsey Lewis[2], Seiya Saito[3], and Marc Fuchs[1]

[1]Department of Plant Pathology and Plant-Microbe Biology, Cornell University, New York State Agricultural Experiment Station, Geneva, NY 14456, USA

[2]Department of Crop Science, North Carolina State University, 4310 Williams Hall Campus, Box 7620, Raleigh, NC 27695-7620, USA

[3]Kearney Agricultural Research Center, University of California, Parlier, CA 93648, USA

ABSTRACT

Background

The relationship between allopolyploidy and plant virus resistance is poorly understood. To determine the relationship of plant evolutionary history and basal virus resistance, a panel of *Nicotiana* species from diverse geographic regions and ploidy levels was assessed for resistance to non-coevolved viruses from the genus *Nepovirus*, family *Secoviridae*. The heritability of resistance was tested in a panel of synthetic allopolyploids. Leaves of different positions on each inoculated plant were tested for virus presence and a subset of plants was re-inoculated and assessed for systemic recovery.

Results

Depending on the host-virus combination, plants displayed immunity, susceptibility or intermediate levels of resistance. Synthetic allopolyploids showed an incompletely dominant resistance phenotype and manifested systemic recovery. Plant ploidy was weakly negatively correlated with virus resistance in *Nicotiana* species, but this trend did not hold when synthetic allopolyploids were taken into account. Furthermore, a relationship between resistance and geographical origin was observed.

Conclusion

The gradients of resistance and virulence corresponded to a modified matching allele model of resistance. Intermediate resistance responses of allopolyploids corresponded with a model of multi-allelic additive resistance. The variable virus resistance of extant allopolyploids suggested that selection-based mechanisms surpass ploidy with respect to evolution of basal resistance to viruses.

BACKGROUND

The 'Red Queen Hypothesis' suggests that coevolution between hosts and pathogens or pests results in a 'boom and bust' cycle where neither host nor its invader can gain lasting supremacy[1]. Allopolyploidy could provide an opportunity for host species to outpace Red Queen coevolution and achieve epochal gains in resistance such as when two moderately-resistant diploids give rise to an allotetraploid with a full complement of resistance genes. This allopolyploid resistance hypothesis incorporates resistance into models explaining heterosis [2,3], and has been tested experimentally in multiple plant and animal systems [4,5]. Allopolyploidization contributes to 2-4% of speciation events in Angiosperms [6].

Viruses have challenged plants for millennia [7-9]. The genus *Nicotiana* has been used as a model system for studying plant-virus interactions and for investigating genotypic and phenotypic changes that occur at and after polyploidization [10]. The genus *Nicotiana* has 76 recognized species, 35 of which are allotetraploids arising from at least five independent interspecific allopolyploidization events [10]. The most likely diploid progenitors of most *Nicotiana* allopolyploids have been determined using nuclear and plastid DNA sequence information [11-14]. While the majority of *Nicotiana* allopolyploids retained their original chromosome number, most species in section *Suaveolentes* underwent a reduction in chromosome number. Genomic changes can occur in the earliest generations following polyploidization [15-18], and all well-studied *Nicotiana*allotetraploids have undergone gene loss or conversion [12,13]. The main center of diversity for*Nicotiana* is Bolivia and the natural range of this genus extends throughout South America, to the Western US, Australia and Africa [10]. In particular, *N. tabacum* and *N. rustica* likely originated in South America, *N. clevelandii* and *N. quadrivalvis* are endemic to the Western US, and all but one species of section *Suaveolentes* are endemic to Australia [10].

Plant viruses are commonly characterized by their experimental host ranges, sometimes incorporating reactions on *Nicotiana* species in their descriptions [19]. The susceptibility status of*N. tabacum* is known for 541 plant viruses, and at least 29 *Nicotiana* species have been used in virus host range studies [20]. Members of *Nicotiana* section *Suaveolentes* (such as *N. benthamiana*) tend to have the widest experimental host ranges [21-23] and *N. benthamiana*'s multi-pathogen susceptibility makes it an important tool for phytopathology research [10,24]. Although the biological basis of *Nicotiana* nonhost resistance to viruses is unknown, a mutated form of *RNA-dependent RNA polymerase 1* in *N. benthamiana* compromises its broad-spectrum antiviral resistance response [25]. Several dominant, strain-specific virus resistance mechanisms have been described in *Nicotiana*[26-28], and closely related viruses exhibit differential capacities for *Nicotiana* systemic infection [27,28].

Interspecific hybridization can be a useful tool for transferring resistance genes to crops species and for investigating virus resistance [19,29,30]. Interspecific (euploid) hybrids of *Solanum tuberosum* and *S. brevidens* showed quantitative resistance to three diverse potato viruses compared to *S. tuberosum,* which exhibited high virus titers after inoculation [31]. The broad-spectrum virus resistance was quantitatively enhanced if the hybrid contained additional copies of the *S. brevidens* genome or if the plants were aneuploids missing an *S. tuberosum* chromosome [31]. Introgression of an alien chromosome from *N. africana* into *N. tabacum* produced tolerance (an amelioration of symptoms) to *Potato virus Y* in *N. tabacum,* but did not confer the immunity exhibited by *N. africana per se*[32]. These data support the conclusion that basal virus resistance is quantitatively controlled by multiple genes.

Nepoviruses are *ne*matode-transmitted *po*lyhedral-shaped viruses of the family *Secoviridae*[33]. These viruses, including *Grapevine fanleaf virus* (GFLV) and *Tomato ringspot virus* (ToRSV), have single-stranded bipartite RNA genomes in positive-sense orientation. GFLV and ToRSV are present in most arable temperate regions and cause severe economic losses to grapevine and other crops [34,35]. Based on the distribution of their highly specific nematode vectors, the likely origins of GFLV and ToRSV are the Near East and Eastern North America, respectively [36,37]. *N. tabacum*exhibits a recovery reaction after infection of GFLV and ToRSV, and salicylic acid (SA)-based resistance mechanisms appear to be critical for recovery from ToRSV [38,39]. RNA silencing mediates *N. tabacum* resistance [40,41] and tolerance [25,42-44] to the nepoviruses *Tomato black ring virus* and *Tobacco ringspot virus*. Although RNA silencing- and SA-based mechanisms of nepovirus resistance have been described, no nepovirus resistance genes have been identified including in well-studied *Vitis* spp. [45], and the diversity and heritability of nepovirus resistance responses are unknown.

Although experimental work has shed light on the effect of allopolyploidy on pest resistance [4,5], very little is currently known about how allopolyploidy could impact evolution of plant virus resistance. The objective of this research was to investigate the relationship between allopolyploidy, geographical origin and genomic bases of basal antiviral responses in *Nicotiana*. The *Nicotiana*-nepovirus pathosystem is a logical choice to test basal (nonspecific) antiviral responses because *Nicotiana* species are generally inbreeding [10], nepovirus strains are genetically stable [46] and these plants and viruses have not coevolved. In this study, we tested the nepovirus resistance status of *Nicotiana* and ascertained heritability using synthetic allopolyploids. We also tested whether the resistance is local or systemically acquired. The central hypotheses were that greater or lesser basal resistance could be explained by geography and ancestry, and that allopolyploids exhibit greater levels of virus resistance than diploids.

RESULTS

Test for Virus Presence

Twenty-four *Nicotiana* species and synthetic allopolyploids of distinct geographic origins were evaluated for their reaction to infection with GFLV strains GHu and F13, and ToRSV strain AP (Table 1). Since GFLV-GHu displays levels of virulence intermediate to that of GFLV-F13 and ToRSV-AP in most *Nicotiana* species, plants were primarily assessed for resistance to GFLV-GHu. Each plant-virus combination was sampled at three or more time points except when a definite resistance or susceptibility determination could be made in the first or second apical leaf i.e. for GFLV-F13-inoculated 4 × (*N. sylvestris* × *N. tomentosiformis*), 4 × (*N. glutinosa* × *N. tabacum*), 4 × (*N. sylvestris* × *N.otophora*) and 4 × (*N. rustica* × *N. tabacum*) (sampled once), and GFLV-GHu-inoculated 4 × (*N. glutinosa* × *N. tabacum*) and *N. goodspeedii* (sampled twice). All panels were surveyed for virus presence in every plant [populations of four to 32 (median 17) plants], except for GFLV-F13-inoculated 2 × (*N. tabacum* × *N. benthamiana*), where 23 plants in an original population of 70 plants was sampled for virus presence in apical leaves in a stratified sampling approach.

Table 1: Sources of *Nicotiana* species and synthetic allopolyploids used in this study

Nicotiana sp.	Authority	Accessiona	Germplasm source	PI number	Originb	Providerc
attenuata	Torr. ex S. Watson	N/A	Bureau of Land Management	W6 27220	SW US Pinyon forest	NCSU
goodspeedii	H.-M. Wheeler	25-G	USDA ARS Beltsville	NSL 8663	Australia	NCSU
obtusifolia	M. Martens and Galeotti	TW98	USDA ARS Beltsville	555543	SW US/ NW Mex	NCSU
debneyi	Domin.	TW36	N/A	N/A	Australia	NCSU
kawakamii	Y. Ohashi	TW72	Iwata Tobacco Experiment Station	459106	Bolivia	NCSU
otophora	Griseb.	TW97	Servicio Agricola Inter-Americano	302477	Ibanex Province, Bolivia	NCSU
paniculata	L.	TW100	C. Rick, Univ Calif	241769	Peru	NCSU
setchelii	Goodsp.	TW121	USDA ARS Beltsville	555557	Peru	NCSU
suaveolens	Lehm.	TW128	CSIRO	230960	Australia	NCSU

tomentosiformis	Goodsp.	TW142	USDA ARS Beltsville	555572	Bolivia	NCSU
glauca	Graham	N/A	World Seed Supply, Mastic Beach, NY	N/A	Bolivia or Argentina	commercial source
glutinosa	L.	N/A	World Seed Supply, Mastic Beach, NY	N/A	Bolivia, Ecuador, Peru	commercial source
rustica	L.	N/A	World Seed Supply, Mastic Beach, NY	N/A	Bolivia, Ecuador or Peru	commercial source
sylvestris	Speg. and Comes	N/A	Botanical Interests, Inc.; Broomfield, CO	N/A	Bolivia or Argentina	commercial source
benthamiana	Domin.	N/A	N/A	N/A	Australia	R. Provvidenti; Cornell
clevelandii	A. Gray	N/A	N/A	N/A	SW US or NW Mexico	R. Provvidenti; Cornell
tabacum cv. Xanthi	L.	N/A	N/A	N/A	Domesticated	R. Provvidenti; Cornell

Hybrid maternal parent	Hybrid paternal parent	PI number	Hybrid accession reference	ploidy	Creator	Donor
sylvestris	tomentosiformis	555722	TH37	amphidiploid	L. Burk; Prosser, WA	NCSU
rustica var. brasilia	tabacum cv. Boltons special	555701	TH34	amphidiploid	Anon.	NCSU
sylvestris	otophora	555721	TH32	amphidiploid	L. Burk; Prosser, WA	NCSU
quadrivalvis	tabacum cv. Red Russian	555515	TH1	amphidiploid	USDA ARS Beltsville	NCSU
glutinosa	tabacum cv. Red Russian	555520	TH10	amphidiploid	USDA ARS Beltsville	NCSU
debneyi	clevelandii	555699	TH15	amphidiploid	Cameron, UC Berkeley	NCSU
tabacum cv. Turkish Sam. (nn) S9-7	benthamiana	N/A	hybrid 230	amphihaploid	G.B. Collins; Lexington, KY	Kentucky State U.

[a]N/A: not available.

[b]Specific origin of the accession is given, where known. Otherwise, the endemic range of the species according to [10] is listed.

[c]NCSU: North Carolina State University.

Gottula *et al.*

DAS-ELISA was used to determine virus presence or absence for 2719 GFLV samples and 536 ToRSV samples in 48 plant-virus combinations. DAS-ELISA absorbance values had a bimodal distribution, which allowed a clear delineation of virus-positive from virus-negative samples. Infection frequencies at each leaf position in each virus-host sample group were summed to calculate virus incidence, which was the dependent variable in correlation analyses. The trajectory of virus incidence among leaf positions in given host-virus combinations were evaluated to generate six discrete resistance categories. Based on the spatial distribution of virus in host plants, host-virus interactions were labeled 'susceptible', 'immune' or one of four categories of recovery ('early', 'intermediate' and late'), or 'delayed susceptibility'.

Symptoms

Virus-inoculated plants were checked regularly for symptoms. The only instances of visible symptoms were for GFLV-GHu on *N. benthamiana, N. clevelandii, N. goodspeedii* and 2 × (*N. tabacum* × *N. benthamiana*), and for ToRSV-AP on *N. benthamiana* and 2 × (*N. tabacum* × *N. benthamiana*). GFLV-GHu symptoms on *N. benthamiana* and *N. clevelandii* were consistent with those previously described [47], and included vein clearing on *N. benthamiana* and amorphous ring-like mottling on *N. clevelandii*. GFLV-GHu symptoms on *N. goodspeedii* included vein clearing analogous to that observed for *N. benthamiana*. GFLV-GHu symptoms on the 2 × (*N. tabacum* × *N. benthamiana*) were composed of non-necrotic ringspots on the first or second leaf position. ToRSV-AP symptoms on *N. benthamiana* were similar to those previously described [48], and included stunting, severe mottling, and necrosis from which the plant ultimately recovered. ToRSV-AP caused mild mottling and slight stunting on 2 × (*N. tabacum* × *N. benthamiana*) and symptoms were not observed on *N. tabacum* cv. Xanthi.

Inoculated Leaf Infection

DAS-ELISA revealed different frequencies of virus infection in inoculated leaves (Figure 1). Some host-virus combinations consistently produced absorbance values below the virus detection threshold, which reflects immunity or perhaps limited subliminal (single cell) infections. 4 × (*N.

sylvestris × *N. tomentosiformis*), 4 × (*N. sylvestris* × *N. otophora*), 4 × (*N. glutinosa* × *N. tabacum*) and *N. paniculata* exhibited immunity to GFLV-F13 in inoculated leaves. Some host-virus combinations resulted in less than 50% inoculated leaf infection including GFLV-GHu-inoculated *N. obtusifolia* (13%) and *N. glauca* (14%), and GFLV-F13-inoculated 4 × (*N. rustica* × *N. tabacum*) (43%) and 4 × (*N. glutinosa* × *N. tabacum*) (44%) . All other tested host-virus combinations produced 50% or greater inoculated leaf infection. Since GFLV-GHu always produced infections in inoculated or apical leaves, and ToRSV-AP inoculations always produced some frequency of infection in the first apical leaf, there is no immunity within this *Nicotiana* panel to these two viruses (Table 2).

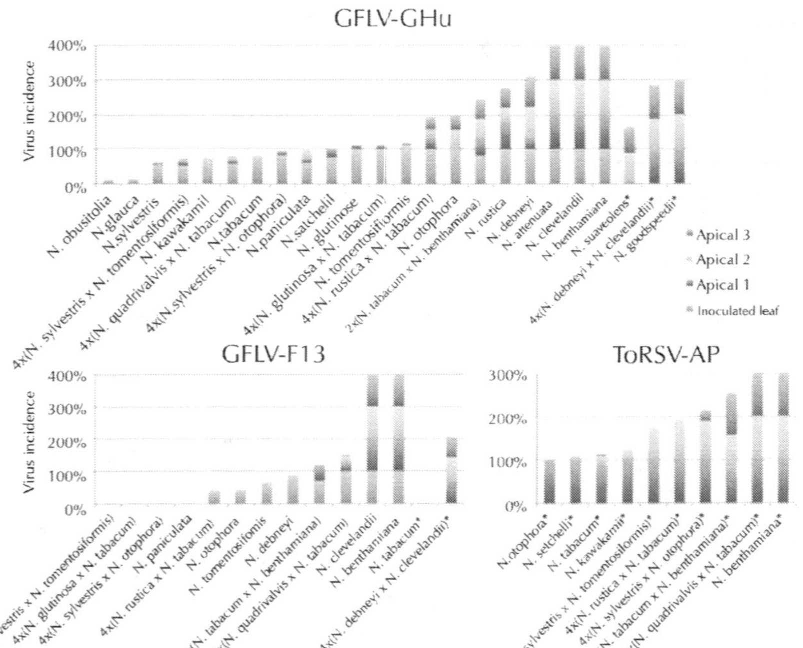

Figure 1: Infection frequencies in inoculated and apical leaves of populations of plants tested for resistance to *Grapevine fanleaf virus* (GFLV) strains GHu and F13, and *Tomato ringspot virus* (ToRSV) strain AP. The percent virus incidence is indicated for inoculated and apical leaves. Asterisks (*) after plant names indicate that the inoculated leaves in the plant-virus combination were not tested.

Table 2: *Grapevine fanleaf virus* and *Tomato ringspot virus* resistance ratings of *Nicotiana* species and synthetic allopolyploids

Sectiona	Origin	x=	Nicotiana species	GFLV-GHu b	GFLV-F13c	ToRSV-AP
Tomentosae	Bolivia	12	N. otophora	5	2	3
	Bolivia	12	N. tomentosiformis	2	2	
	Bolivia	12	N. kawakamii	2		3
	Peru	12	N. setchelii	3		3
Paniculatae	Peru	12	N. paniculata	3	1	
Noctiflorae	Bolivia	12	N. glauca	2		
Sylvestres	Bolivia	12	N. sylvestris	2		
Undulatae	Peru	12	N. glutinosa	3		
Trigonophyllae	SW US	12	N. obtusifolia	2		
Petunioides	SW US	12	N. attenuata	6		
Suaveolentes	Australia	16	N. suaveolens	5		
	Australia	19	N. benthamiana	6	6	6
	Australia	20	N. goodspeedii	6		
	Australia	24	N. debneyi	4	2	
Polydicliae	SW US	24	N. clevelandii	6	6	
Rusticae	Bolivia	24	N. rustica	4		
Nicotiana	Domesticated	24	N. tabacum	2	1 or 2	3
Wide crosses (synthetic allopolyploids)		24	4x(N. sylvestris x N. otophora)	3	1	3
		24	4x(N. sylvestris x N. tomentosiformis)	3	1	3
		36	4x(N. glutinosa x N. tabacum)	3	1	
		43	2x(N. tabacum x N. benthamiana)	4	3	5
		48	4x(N. quadrivalvis x N. tabacum)	3	3	6
		48	4x(N. rustica x N. tabacum)	4	2	3
		48	4x(N. debneyi x N. clevelandii)	5	4	

[a]Species and synthetic allopolyploids are referenced by their sections within the genus *Nicotiana*, primary location of origin, and their haploid chromosome numbers according to [10].

[b]Categories of resistance (1, most resistant, through 6, most susceptible) are indicated for each virus-host combination tested.

[c]Boxes without resistance ratings represent virus-host combination not tested.

Gottula *et al.*

High Resistance Interactions

Virus-host combinations yielding no detectable virus in inoculated leaves (and apical leaves) were designated as immune (category 1). Immunity was observed for *N. paniculata,* 4 × (*N. sylvestris* ×*N. tomentosiformis*), 4 × (*N. sylvestris* × *N. otophora*) and 4 × (*N. glutinosa* × *N. tabacum*) inoculated with GFLV-F13 (Figure 1; Table 2). GFLV-F13-inoculated *N. tabacum* did not produce apical leaf infection, but whether this plant is immune (category 1) or possesses early recovery (category 2) to GFLV-F13 is uncertain because inoculated leaves were not tested. All tested members of section *Tomentosae, N. debneyi* and 4 × (*N. rustica* × *N. tabacum*) exhibited early recovery (category 2) to GFLV-F13 (Table 2). *N. obtusifolia, N. glauca, N. sylvestris, N. kawakamii,N. tabacum* and *N. tomentosiformis* exhibited early recovery after GFLV-GHu inoculation. Early recovery was not observed for these species in response to inoculation with ToRSV-AP (Table 2).

Moderate Resistance Interactions

Late recovery (category 3) was the most frequent host-virus interaction phenomenon observed in this test panel, and was seen for all virus isolates tested. All tested members of section*Tomentosae,* 4×(*N. rustica* × *N. tabacum*), *N. tabacum* and resynthesized allopolyploids involving possible *N. tabacum* progenitor species [4 × (*N. sylvestris* × *N. tomentosiformis*) and 4 × (*N. sylvestris* × *N. otophora*)] showed late recovery to ToRSV-AP (Table 2). 2 × (*N. tabacum* × *N. benthamiana*) and 4 × (*N. quadrivalvis* × *N. tabacum*) showed late recovery to GFLV-F13, and 4x(*N. quadrivalvis* × *N. tabacum*), 4 × (*N. sylvestris* × *N. otophora*), 4 × (*N. sylvestris* × *N. tomentosiformis*), 4 × (*N. glutinosa* x *N. tabacum*), *N. glutinosa, N. paniculata* and *N. setchelii*showed late recovery to GFLV-GHu (Table 2). Intermediate recovery (category 4), characterized by fluctuation of virus incidence over three or more leaf axes (typically between 33% and 67%, was observed in GFLV-GHu-inoculated *N.*

debneyi, N. rustica, 4 × (*N. rustica* × *N. tabacum*) and 2 × (*N. tabacum* × *N. benthamiana*), and in GFLV-F13-inoculated 4 × (*N. debneyi* × *N. clevelandii*) (Table 2).

Low or No Resistance Interactions

Delayed susceptibility (category 5) was observed only in response to GFLV-GHu inoculation of *N. otophora, N. suaveolens,* and 4 × (*N. debneyi* × *N. clevelandii*) (Figure 1; Table 2). Plants were designated as susceptible (category 6) when 100% of the plants became infected and virus was present in all tested leaves. *N. benthamiana* and *N. clevelandii* were susceptible to GFLV-F13 and GFLV-GHu, as expected [47], *N. goodspeedii* and *N. attenuata* were susceptible to GFLV-GHu, and*N. benthamiana* and 4 × (*N. quadrivalvis* × *N. tabacum*) were susceptible to ToRSV-AP (Figure 1; Table 2).

Additive Resistance Phenomena in Synthetic Polyploid Plants

Incompletely dominant virus resistance was observed in synthetic *Nicotiana* allopolyploids. Whereas *N. tabacum* showed high resistance to GFLV-GHu, ToRSV-AP and GFLV-F13, and *N. benthamiana* was fully susceptible to all three virus strains, 2 × (*N. tabacum* × *N. benthamiana*) exhibited delayed susceptibility to GFLV-GHu, intermediate recovery to ToRSV-AP, and late recovery to GFLV-F13 (Figure 2; Table 2). Whereas *N. clevelandii* was fully susceptible to all viruses tested, and *N. debneyi* exhibited early recovery to GFLV-F13 and intermediate recovery to GFLV-GHu, 4 × (*N. debneyi* × *N. clevelandii*) exhibited intermediate recovery to GFLV-F13 and delayed susceptibility to GFLV-GHu (Figure 2, Table 2). The 4 × (*N. rustica* × *N. tabacum*) response to GFLV-GHu was not categorically different than the response of *N. rustica* (both category 4), but the synthetic allopolyploid showed consistently lower incidence of infection in apical leaves (23-40%) compared to *N. rustica* (40-80%), which could reflect the contribution of *N. tabacum*(category 2) to resistance (Figure 1). The intermediate virus resistance observed across *Nicotiana* lineages (Figure 3) suggests quantitative resistance is not due to a single gene with dosage effects, but due to multiple genes with dosage effects.

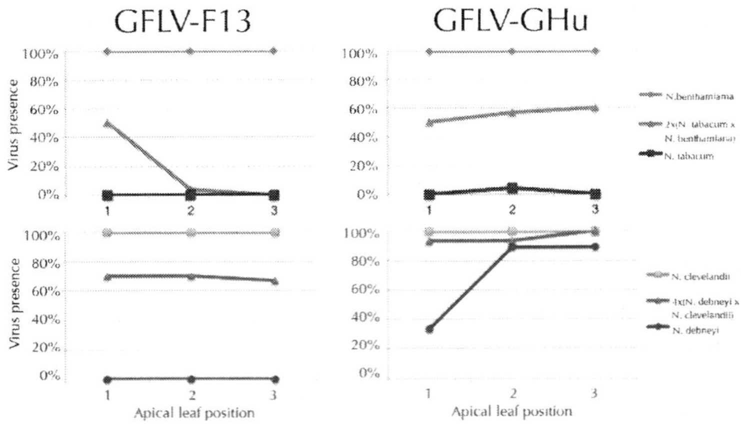

Figure 2: Effect of synthetic *Nicotiana* allopolyploids on resistance to *Grapevine fanleaf virus* (GFLV) strains F13 (left panels) and GHu (right panels). *N. tabacum, N. benthamiana* and the 2x(*N. tabacum* x *N. benthamiana*) amphihaploid (upper panels); and *N. debneyi, N. clevelandii* and 4x(*N. debneyi* x *N. clevelandii*) allopolyploid (lower panels) were tested for additive resistance.

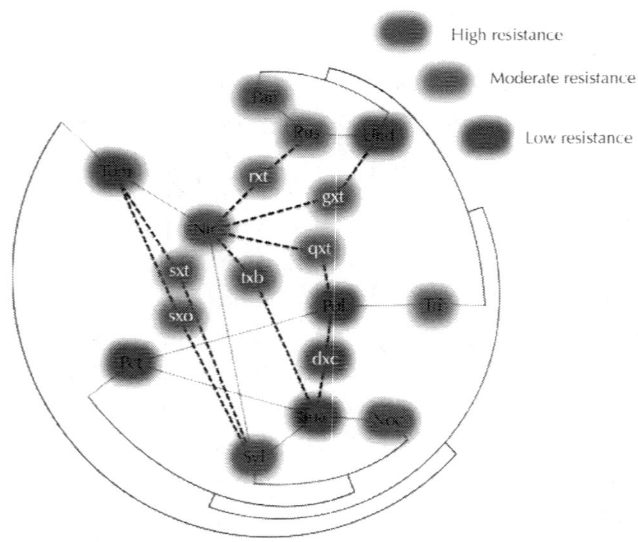

Figure 3: *Grapevine fanleaf virus* strain GHu (GFLV-GHu) resistance categories superimposed on a *Nicotiana* phylogenetic tree modified from

Clarkson *et al.*[18](curved lines), containing sections (abbreviated in black lettering) with allopolyploid ancestries as established by Clarkson *et al.* [13]and Kelly *et al.*[14](solid straight lines). Shading surrounding sections denote the resistance category of representative species tested for GFLV-GHu resistance: blue (category 2, early recovery), purple (categories 3 and 4, late or intermediate recovery), or red (categories 5 and 6, delayed or full susceptibility). Representative *Nicotiana* species (sections) used in this study are *N. paniculata* (*Paniculatae, 'Pan'*), *N. rustica* (*Rusticae, 'Rus'*), *N. obtusifolia* (*Trigonophyllae, 'Tri'*), *N. benthamiana, N. debneyi, N. suaveolens* and *N. goodspeedii* (*Suaveolentes, 'Sua'*), *N. clevelandii* (*Polydicliae, 'Pol'*), *N. glauca* (*Noctiflorae, 'Noc'*), *N. sylvestris* (*Sylvestres 'Syl'*), *N. tabacum* (*Nicotiana 'Nic'*), *N. glutinosa* (*Undulatae, 'Und'*) *N. attenuata* (*Petunioides, 'Pet'*), (*Tomentosae, 'Tom'*) including *N. kawakamii, N. otophora, N. setchelii* and *N. tomentosiformis*. Members of *Tomentosae* and *Suaveolentes*exhibited different GFLV-GHu resistance profiles and are accordingly dually or triply colored. Descent of synthetic allopolyploids used in this study (white letters) is indicated by dashed lines: *4*x(*N. sylvestris* x *N. tomentosiformis*) ('sxt'), *4*x(*N. rustica* x *N. tabacum*) ('rxt'), *4*x(*N. glutinosa* x *N. tabacum*) ('gxt'), *2*x(*N. tabacum* x *N. benthamiana*) ('txb'), *4*x(*N. quadrivalvis* x*N. tabacum*) ('qxt') and *4*x(*N. debneyi* x *N. clevelandii)* ('dxc').

Resistance Profiles of Allopolyploids and their Progenitors

We tested two natural allopolyploids (*N. clevelandii* and *N. tabacum*) and the closest relatives of their known progenitors for GFLV-GHu resistance. The closest extant diploid progenitors of *N. clevelandii* are *N. obtusifolia* (maternal genome donor) and *N. attenuata* (paternal genome donor) [13]. While *N. obtusifolia* exhibited an early recovery phenotype (category 2), both *N. clevelandii*and *N. attenuata* showed complete susceptibility (Figure 1;Table 2). *N. tabacum*, its representative maternal genome donor (*N. sylvestris*), and one possible representative paternal genome donor (*N. tomentosiformis*) each exhibited early recovery (category 2). *N. otophora*, another representative of *N. tabacum's* possible paternal genome donors exhibited delayed susceptibility (category 5) to GFLV-GHu. Resynthesized allopolyploids corresponding to either *N. tabacum*ancestry scenario exhibited late recovery phenotypes (category 3) to GFLV-GHu with low virus incidence levels (Figure 1. Thus, *N. tabacum* exhibits an early recovery phenotype similar to that of its maternal genome donor and of *N. tomentosiformis*, but less resistance than that of *N. otophora* or representative resynthesized allopolyploids. Additionally, species of section *Suaveolentes* showed low or occasionally moderate resistance to GFLV-GHu, while its most closely related proposed paternal genome donor (*N. sylvestris*) [14] showed high resistance (early recovery) (Table 2). While neoallopolyploids showed intermediate GFLV-GHu resistance characteristics, extant allopolyploids did not show intermediate GFLV-GHu resistance characteristics (Figure 3).

Systemic Recovery

Systemic recovery was tested in apical leaves of GFLV-resistant (categories 1 or 3) synthetic allopolyploids 4 × (*N. sylvestris* × *N. tomentosiformis*), 4 × (*N. glutinosa* × *N. tabacum*), and 4 × (*N. sylvestris* × *N. otophora*) (Table 3). Resistance was induced with GFLV-GHu or GFLV-F13, and one upper, apical leaf of each recovered plant was re-inoculated with GFLV-GHu and tested for virus presence. Notably, plants that showed inoculated leaf susceptibility to GFLV-GHu lost this susceptibility in the apical leaf of the recovered plant, no matter whether the resistance was induced with GFLV-F13 or GFLV-GHu (Table 3). GFLV-GHu was occasionally detected in the apical inoculated leaf of GFLV-GHu-recovered plants encompassing two of 21 plants in 4 × (*N. sylvestris* ×*N. tomentosiformis*) and one of nine plants in 4 × (*N. sylvestris* × *N. otophora*). Of the plants that did not acquire systemic recovery, the possibility of the originally-inoculated GFLV-GHu infected these apical leaves cannot be excluded given that late recovery does not bar the virus from infecting the fourth leaf position, albeit at a low incidence.

Table 3: Systemic recovery from *Grapevine fanleaf virus* (GFLV) strains F13 and GHu

		1st inoculationa	Resistance response	Systemic recoveryb
4x(N. sylvestris x N. tomentosiformis)		GFLV-GHu	Late recovery	90%
GFLV-F13		Immunity	100%	

4x(N. glutinosa x N. tabacum)	GFLV-GHu	Late Recovery	100%
GFLV-F13	Immunity	100%	
4x(N. sylvestris x N. otophora)	GFLV-GHu	Late Recovery	89%

[a]GFLV-GHu was inoculated to the fourth apical leaf following induction of resistance (resistance response against the virus in the 1st **inoculation is indicated).**

[b]Plants ($n = 9$ to 29) were characterized as having systemic recovery if GFLV was undetectable at five days post-inoculation.

Gottula *et al.*

Relationship between Host Geographic Origin and Virus Resistance

Australian and North American accessions generally displayed greater levels of susceptibility than South American accessions to all virus strains tested (Table 2). The Australian species *N. benthamiana* and the North American species *N. clevelandii* were fully susceptible to all viruses tested, and Australian species *N. debneyi, N. suaveolens* and *N. goodspeedii*, and North American species *N. attenuata* and *N. quadrivalvis* displayed lower levels of resistance than South American species to GFLV-GHu individually or in hybrid backgrounds (Table 2). Exceptions to these geography-based resistance trends included the *N. debneyi* (Australia) early recovery response to GFLV-F13, the *N. obtusifolia* (North America) early recovery response to GFLV-GHu, and the *N. otophora* (South America) delayed susceptibility response to GFLV-GHu. Overall, origin had a significant ($P < 0.0001$) and moderate correlation for GFLV-GHu incidence when hybrids were excluded from the analysis ($r = 0.683$) and a weaker correlation ($r = 0.5422$, $P < 0.0001$) when hybrids were included, with South American species showing greater resistance than Australian species, which in turn showed greater resistance than species from the Southwest US. Because the effect of section cannot be separated from the effect of origin (Table 2), the effect of origin on virus resistance could reflect phylogenetic factors.

Limited Relationship between Host Ploidy Level and Virus Resistance

There was a weak association between ploidy level and virus susceptibility. For example, $n = 12$ diploids from section *Tomentosae* generally displayed greater levels of resistance than $n = 16$-24 allopolyploids of section *Suaveolentes,* and similar levels of resistance to *N. tabacum* and *N. rustica*($n = 24$) (Table 2). The correlation between GFLV-GHu incidence and chromosome number was low ($r = -0.036$) and nonsignificant ($P = 0.2597$) when hybrids were included in the analysis, and low ($r = -0.286$) but significant ($P < 0.0001$) when hybrids were excluded, indicating that increasing ploidy is weakly negatively related to GFLV-GHu virus incidence among extant *Nicotiana* species. These results indicate that increasing ploidy is correlated with slightly greater virus susceptibility, but that the trend is abolished when synthetic allopolyploids are taken into account.

Other Trends in Virus Resistance

Members of section *Tomentosae* produced higher inoculated leaf infection rates (75-100%) for GFLV-GHu than for GFLV-F13 (44-67%). Every tested member of section *Tomentosae* produced an early recovery phenotype for GFLV-F13 and a late recovery phenotype for ToRSV-AP (Table 2). Members of section *Tomentosae* showed variability in response to GFLV-GHu, where *N. kawakamii* and *N. tomentosiformis* exhibited early recovery, *N. setchelii*displayed late recovery, and *N. otophora* showed delayed susceptibility (Table 2). The delayed susceptibility of *N. otophora* to GFLV-GHu was masked in the 4 × (*N. sylvestris* × *N. otophora*) synthetic allopolyploid, which reflected the early recovery of *N. sylvestris* to GFLV-GHu (Table 2). Early recovery was also observed for *N. tabacum* inoculated with GFLV-GHu, a species believed to have evolved from a *N. sylvestris* × *N. otophora* or *N. sylvestris* × *N. tomentosiformis*hybridization event [10]. Members of section *Suaveolentes* exhibited intermediate or low resistance to the nepovirus strains tested, except for *N. debneyi,* which displayed early recovery after inoculation with GFLV-F13 (category 2) (Table 2).

N. tabacum and its corresponding resynthesized allopolyploids [4 × (*N. sylvestris* × *N. otophora*) and 4 × (*N. sylvestris* × *N.*

tomentosiformis)] exhibited high or moderate virus resistance phenotypes for each virus tested (Table 2). Both resynthesized allopolyploids are immune to GFLV-F13, and *N. tabacum* also displays high resistance to this virus. *N. tabacum* and its resynthesized allopolyploids showed late recovery to ToRSV-AP, though *N. tabacum* frequently had lower frequencies of infection at any given leaf position than its corresponding neoallopolyploids (Figure 1). The response of *N. tabacum* and the synthetic allopolyploids 4 × (*N. sylvestris* × *N. otophora*) and 4 × (*N. sylvestris* × *N. tomentosiformis*) to GFLV-GHu were similar in terms of inoculated leaf infection, but *N. tabacum* showed early recovery whereas the neoallopolyploids showed late recovery, though the overall apical virus incidence levels were similar The recovery responses of *N. tabacum* to GFLV and ToRSV inoculation confirm previous reports [38,39].

Synthetic polyploids formed from resistant and susceptible species frequently displayed resistance in the moderate categories (Figure 3). 2 × (*N. tabacum* × *N. benthamiana*) and 4 × (*N. debneyi* × *N. clevelandii*) exhibited intermediate resistance phenotypes after inoculation with GFLV-GHu and GFLV-F13 compared to their parents (Figure 2; Table 2). The same was true for the 2 × (*N. benthamiana* × *N. tabacum*) response to ToRSV-AP (Table 2). An intermediate level of apical leaf infection was also seen in the 4 × (*N. rustica* × *N. tabacum*) response to GFLV-GHu (Figure 1).

ToRSV-AP typically produced equal or greater categorical ratings than GFLV-GHu, and GFLV-GHu always produced equal or higher category ratings than GFLV-F13 (Table 2). An exception to this virulence trend was that *N. otophora* and 4 × (*N. rustica* × *N. tabacum*) showed lower resistance (higher category ratings) to GFLV-GHu than to ToRSV-AP (Table 2). Virulence differences between GFLV-F13 and GFLV-GHu were highly apparent in synthetic allopolyploid plants with resistant and susceptible parents, including 2 × (*N. tabacum* × *N. benthamiana*), 4 × (*N. rustica* × *N. tabacum*), and 4 × (*N. debneyi* × *N. clevelandii*) (Table 2; Figure 2). There was a significant ($P < 0.0001$) but weak ($r = 0.406$) correlation between virus composition and infection frequencies across plant genotypes (species or synthetic allopolyploids).

According to individual components of ^{2} in the contingency table that compared observed and expected virus incidence frequencies for each virus at each leaf position, there is a higher virus incidence in the first apical leaf than expected for ToRSV-AP; conversely, there is less virus incidence in the first apical leaf than expected for GFLV-F13 (data not shown). Expected and observed apical virus incidence values are similar for GFLV-GHu. These results suggest that ToRSV-AP displays higher virulence and GFLV-F13 displays lower virulence than GFLV-GHu in this panel of *Nicotiana* species.

DISCUSSION

A spectrum of plant resistance and viral virulence was observed in the present *Nicotiana*-nepovirus panel. Changes in virus incidence were characterized using DAS-ELISA on multiple leaves of large samples of plants (Figure 1) and used to distill six categories of host resistance (immunity, susceptibility, and four categories of recovery) from which *Nicotiana* species, synthetic allopolyploids and viruses were compared. While all host-virus combinations exhibiting low leaf inoculation frequencies (<50%) exhibited early recovery, this phenotype was frequently associated with a high infection frequency (>50%) in inoculated leaves (Figure 1). Moderate or high leaf inoculation frequencies (≥50%) were associated with an entire range of resistance and susceptibility phenotypes (category 2 through category 6) (Figure 1). Within individual plant genotypes, ToRSV-AP generally produced higher susceptibility ratings than GFLV-GHu, and GFLV-GHu always produced an equal or greater susceptibility rating than GFLV-F13 (Table 2), and the correlation between virus identity and virus incidence ratings were significant. The spectra of quantitative resistance displayed by *Nicotiana* accessions and virulence among nepoviruses suggest the role of multiple interacting alleles from *Nicotiana* accessions and nepoviruses in the determination of the ultimate infection outcomes. Similar plant genotype by virus genotype interactions were observed in a panel of 21 *Arabidopsis* accessions challenged with three*Cucumber mosaic virus* isolates [49].

The full susceptibility seen for 4 × (*N. quadrivalvis* × *N. tabacum*) and delayed susceptibility of the 2 × (*N. tabacum* × *N. benthamiana*) responses to ToRSV raises the interesting possibility that *N. quadrivalvis* and *N. benthamiana* may possess a dominant ToRSV susceptibility factor in *N. tabacum* backgrounds. The observation of ringspot symptoms on the GFLV-GHu-inoculated 2 × (*N. tabacum* x *N. benthamiana*) amphihaploid suggests that the vein clearing symptomology typical of *N. benthamiana* infection [47] is a recessive trait. Similarly, while ToRSV-AP produced necrosis on *N. benthamiana,* necrosis was not observed on the 2 × (*N. tabacum* × *N. benthamiana*) amphihaploid or on *N. tabacum*. The absence of *N. tabacum*-ToRSV necrotic ringspot symptoms was unexpected given previous reports [39,50]. The lack of hypersensitive responses observed in this host panel is consistent with the lack of involvement of a specific gene-for-gene recognition system in *Nicotiana*-GFLV and *Nicotiana*-ToRSV interactions. This lack of hypersensitive response and the absence of coevolutionary history between *Nicotiana* and GFLV or ToRSV supports the idea that resistance or susceptibility is due to the interaction of broad-spectrum immune responses and virulence factors [51].

Most plants in the host panel used in this study recovered from virus infection after infection was initially established in inoculated leaves. Recovery from virus infection can be controlled by simple or complex host plant genetics, and can be countered by effective pathogen virulence factors [30,52,53]. Host plant and pathogen genotype determined the level of plant recovery to GFLV (Figure 2). Compatibility between host and viral components is a prerequisite for infection in the matching allele model [54,55]. The partial resistance phenotypes observed in this study do not fit with the strict bimodality of the matching allele concept. However, a modified matching allele model that allows for partial compatibility and limited infection [1,55] (Figure 4) could explain the range of resistance and virulence observed in the *Nicotiana*-nepovirus interactions observed here.

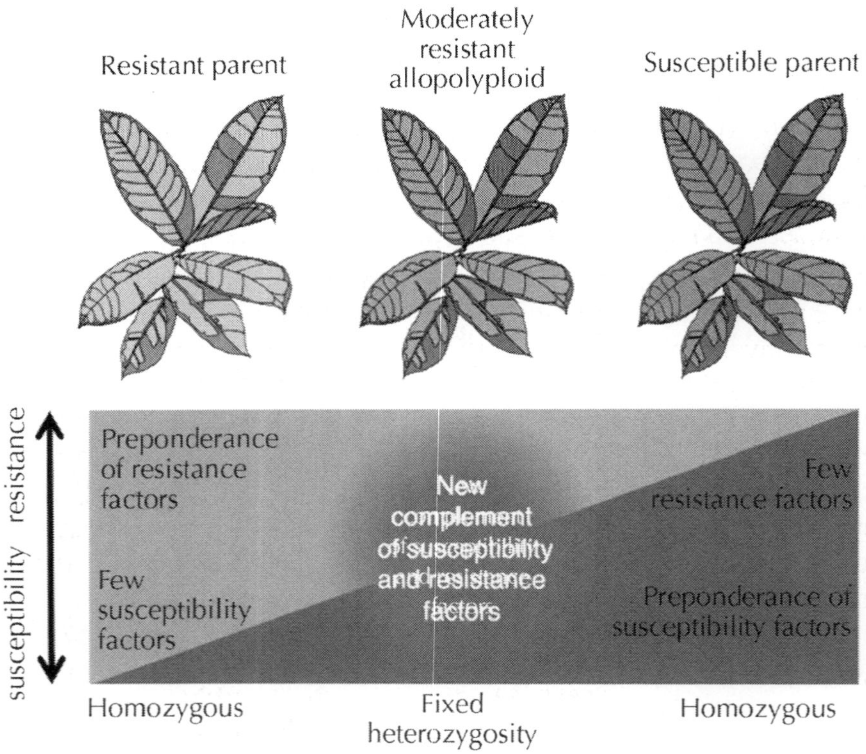

Figure 4: Pictographic description of the modified matching allele model applied to the additive resistance hypothesis.Resistant and susceptible parents (e.g. diploid progenitors of an allopolyploid) carry unique complements of resistance factors (blue) and susceptibility factors (red). The allopolyploid plant would maintain a mix of resistance and susceptibility factors from each parent (fixed heterozygosity), and also would be expected to exhibit unique (nonadditive) expression profiles of resistance and susceptibility factors.

The intermediate resistance responses of *Nicotiana* neoallopolyploids are congruent with the additive resistance hypothesis proposed by Fritz *et al.* [56]. By applying the modified matching allele model to the additive resistance hypothesis, we theorize that susceptible parents contribute susceptibility alleles and resistant parents contribute resistance alleles, and their neoallopolyploids contain novel combinations of resistance and susceptibility factors (Figure 4). Neoallopolyploids would possess a greater number of matching alleles than their more resistant parent, but the dosage of resistance factors would be reduced compared to the resistant parent. Furthermore, non-additive gene expression, which is commonly observed in allopolyploids and other hybrids[57-59], could modify expression of resistance and susceptibility alleles (Figure 4).

While the identities of the *Nicotiana*'s nepovirus resistance alleles are unknown, re-inoculation experiments (Table 3) show that the resistance signal is translocated to result in systemic recovery. Because the *N. tabacum* ToRSV resistance response appears to be SA-mediated [39], susceptibility alleles conferred by *N. benthamiana* in the 2 × (*N. tabacum* × *N. benthamiana*) hybrid could allow ToRSV to quantitatively inhibit SA biosynthesis, affect conversion of SA to an alternate derivative, or vitiate downstream SA-activated resistance responses [60,61]. Similarly, null or ineffective RNA silencing alleles present in susceptible backgrounds could conceivably compromise RNA silencing-mediated virus resistance in hybrids [62]. These hypotheses are consistent with Fraser's model of virus resistance [29], which postulates that the effects of resistance alleles are proportional to their dosage and levels of influence on resistance pathways.

Although animal allopolyploids frequently show dominant parasite susceptibility [4,5,63], *Nicotiana*neoallopolyploids exhibit virus resistance greater than one but not both of their parents (Figure 3). In cases where both parents were either resistant or susceptible, the neoallopolyploid displayed a resistance response similar to their parents, and thus there was no inherent penalty or benefit from hybridization or genome duplication (Figure 3). Contrary to the model that neoallopolyploid plants could face a depression of innate immunity [64], our findings suggest that allopolyploidization itself did not penalize *Nicotiana* for virus resistance.

'Revolutionary changes' that accompany polyploidy can be distinguished from 'evolutionary changes,' which follow allopolyploidization [65,66]. The maintenance of virus resistance in *N. tabacum* contrasts with the apparent loss of virus resistance in section *Polydicliae*, which did not maintain partial virus resistance imparted by its likely maternal genome donor (*N. obtusifolia*) (Figure 3). Similarly, members of *Suaveolentes* exhibited high degrees of virus susceptibility despite the resistance of their paternal genome donor's closest relative (*N. sylvestris*). Low virus resistance in sections *Polydicliae* and *Suaveolentes* suggests genetic drift and/or selection conferred a loss of virus resistance inherited by neoallopolyploids.

Nicotiana neoallopolyploids show gene loss and neofunctionalization [12,16,18]. Since favorable alleles have a lower chance of becoming fixed in allopolyploids than diploids [6], drift could have resulted in losses of innate immunity alleles in the *Polydicliae* and *Suaveolentes* lineages (Figure 5).

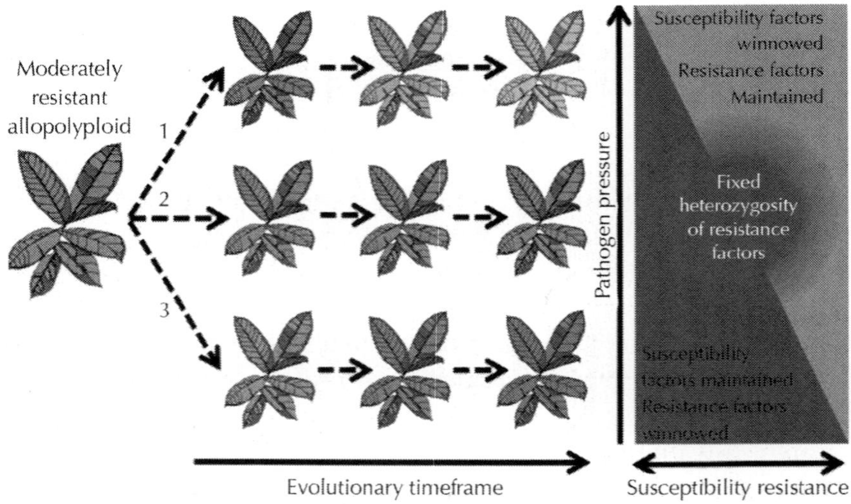

Figure 5: Model of changes in quantitative innate virus resistance from a moderately resistant progenitor exhibiting fixed heterozygosity for resistance genes (e.g. a neoallopolyploid). Random divergence of the allopolyploid progeny leads to several possible lineages containing different resistance or susceptibility phenotypes whose existence depends on drift and pathogen pressure. High pathogen pressure would select for the loss of susceptibility factors and maintenance and gain of resistance factors (trajectory 1, top row). Moderate or irregular pathogen pressure would maintain an equilibrium of resistance and susceptibility factors within the plant population (trajectory 2, middle row). Low pathogen pressure would remove the selective advantage of maintaining resistance factors, and could result in the loss of resistance factors and the maintenance of susceptibility factors (trajectory 3, bottom row).

The correlation of higher levels of virus resistance in South American *Nicotiana* species than North American and Australian species suggest that geographic influences had a major effect on the efficacy of antiviral resistance responses. Alternatively, because plant taxon (section) is frequently inseparable from origin, there is a possibility that phylogeny rather than origin could account for virus resistance. In either scenario, long-term biota-specific interactions would be critical factors to select for improved virus resistance. Existing virus resistance alleles could be maintained or enhanced if virus challengers perennially recur (trajectories 1 or 2), or virus resistance alleles could be lost if virus challenges diminish (trajectory 3) (Figure 5). Comparison of several allopolyploids used in these virus resistance experiments suggest that *N. tabacum*, endemic to the relatively large and competitive biome of the Eastern Andes has followed trajectory 1, while members of sections *Suaveolentes* and *Polydicliae*, endemic to the relatively isolated biota of Australia and Southwest US, respectively, have followed trajectories 2 or 3 (Figure 5). Because *N. debneyi* and *N. benthamiana* are monophyletic [10], but *N. debneyi* shows greater antiviral resistance than *N. benthamiana* (Table 2), random or selection-based processes may have driven divergence of innate immune functions within this allopolyploid lineage. The sister allopolyploids *N. clevelandii* and *N. quadrivalvis* have similarly diverged for herbivory resistance responses [67].

Changes in immune function due to allopolyploidy could precipitate changes in challenging pathogens, and prompt a Red Queen-type evolutionary response between the plant and pathogen[5]. Ineffective innate immune systems could allow otherwise ill adapted viruses to acquire more effective virulence factors and erode quantitative resistance [68,69]. An allopolyploid that can endure colonization by a pathogen or pest and that permits pathogen adaptation to an otherwise resistant host progenitor is referred to as a hybrid bridge [70]. Rather than escaping virus infection, allopolyploids could furnish another niche for viruses to expand their host ranges.

CONCLUSIONS

The Red Queen Hypothesis explains how resistance and virulence temporally change in parasitic relationships. Allopolyploidy might represent an opportunity for plant hosts to break the Red Queen cycle of coevolution by gaining a new complement of dominant resistance factors, but the potential for allopolyploids to experience an epochal gain in innate immune function may be compromised by the inheritance of susceptibility alleles or genetic dysfunctionality caused by hybridization.

TheNicotiana-nepovirus interaction sheds light on potential dynamics of how allopolyploidy may affect innate immunity. Based on a detailed survey of the interaction of non-coevolved plant and virus species, it appears that host and pathogen genotypes contain multiple alleles that interact in a quantitative fashion to determine the level of resistance or susceptibility. Synthetic allopolyploids faithfully display additive virus resistance characteristics that correspond to modified matching allele interactions (Figure 4). Virus resistance/susceptibility factors change in allopolyploid progeny due to classical drift and selection (Figure 5). These changes raise the interesting possibility that moderately resistant allopolyploids could provide a hybrid bridge, which could result in a new Red Queen cycle of coevolution.

METHODS

Ethics Statement

This research did not involve research on human subjects, human material or human data, nor did it involve work with regulated invertebrates.

Plant Material

Seventeen *Nicotiana* species and seven synthetic allopolyploids (Table 1) were assayed for nepovirus resistance. With the exception of 2 × (*N. tabacum* × *N. benthamiana*), an infertile amphihaploid, all genetic materials were self-fertile. The synthetic allopolyploids exhibited no obvious phenotypic segregation. Seeds of 2 × (*N. tabacum* × *N. benthamiana*) [71] were a gift from Dr. G.B. Collins's research program (University of Kentucky, Lexington, KY). Seeds of *N. benthamiana, N. tabacum* cv. Xanthi and *N. clevelandii* were from Drs. D. Gonsalves and R. Provvidenti (Cornell University New York State Agricultural Experiment Station, Geneva, NY). Seeds of *N. rustica, N. glauca, N. glutinosa* and *N. sylvestris* were obtained from commercial sources (Table 1). All of the other *Nicotiana* seeds were provided by the United States Nicotiana Germplasm Collection maintained at North Carolina State University (Raleigh, NC). Seedlings were grown in four-inch pots containing soilless potting media. Plants were grown in a greenhouse maintained at 24-26°C supplemented with high pressure sodium lamps for an 18 hour light/ 8 hour dark photoperiod, and watered daily or every other day as needed, and fertilized weekly.

Virus Strains and Inoculation Procedure

GFLV strain F13 from France [72,73] and strain GHu from Hungary [47,74] were isolated from infected grapevines, and ToRSV strain AP was isolated from an infected apricot tree in New York State [75]. GFLV and ToRSV strains were maintained in *N. benthamiana*. Virus inoculum was prepared by mechanically inoculating *N. benthamiana* and storing infected tissue at -80°C until inoculation of the host panel. Infected *N. benthamiana* tissue was ground 1:10 (w:v) in inoculation buffer (15 mM Na_2HPO_4 and 35 mM KH_2PO_4 pH 7.0) using a steel grinding set in a tissue lyser (Qiagen, Valencia, CA) and inoculated to three corundum-dusted leaves of each test plant with a ceramic pestle. Panels of four to 32 (median 17) plants per virus-host combination were selected for uniformity in size and mechanically inoculated when they had 4–5 leaves and were approximately 3 cm in height. All plants were rinsed with water five to ten minutes after inoculation.

Sampling and Virus Tests

Apical leaf positions were defined by counting nodes on the whorl upwards from the highest inoculated leaf. Apical leaves were collected at time points sufficient to detect cumulative virus infection: nine to 18 dpi for position one, 17 to 28 dpi for position two, and 26–60 dpi for position three. In plants where intermediate resistance phenotypes were observed, additional collections were made at 41 to 57 dpi for position four. Inoculated leaves were collected and processed between 21 and 54 dpi.

Plant tissue was collected from inoculated plants and processed for virus detection via double antibody-sandwich enzyme-linked immunosorbent assay (DAS-ELISA). Fresh tissue was ground in 1:10 (w:v) in 25 mM sodium phosphate buffer using a semi-automated HOMEX 6 tissue homogenizer and mesh grinding bags (Bioreba, Reinach, Switzerland). DAS-ELISAs for GFLV and ToRSV were carried out in Nunc MaxiSorp® flat-bottom 96 well polystyrene microtiter plates (Fisher Scientific, Pittsburgh, PA) according to the manufacturer's protocol (Bioreba). Absorbance (OD_{405nm}) was measured after two hours of substrate incubation using a BioTek Synergy2 plate reader and Gen5 software was used to calculate blank-subtracted absorbance (Biotek, Winooski, VT). Each ELISA plate contained positive and negative checks, and the validity of each assay was ascertained before data was processed. Samples were considered positive if their absorbance values were greater than two times the mean absorbance values of negative controls.

Evaluation of Infection Phenotypes

Virus symptoms were monitored daily on inoculated and apical plant leaves. Leaf samples that were positive or negative in DAS-ELISA for GFLV or ToRSV in each inoculation group were counted and converted into percent infection at each leaf position. Six resistance categories were assigned based on the infection outcome in inoculated leaves and in successive apical leaves. Virus-host combinations that yielded no detectable virus in the inoculated leaf (and apical leaves) were designated as 'immune' (category 1). 'Early recovery' (category 2) was defined as any level of inoculated leaf infection (10% to 100%) but the virus was rarely or infrequently (<10%) detected in the first apical leaf. 'Late recovery' (category 3) was defined at 10% to 100% infection in the first or second apical leaf position but a decline in virus incidence at higher leaf axes. 'Intermediate recovery' (category 4) was defined as 20% to 80% infection frequencies in all leaf axes, and no clear pattern of reduction or expansion of virus incidence in successively higher axes. 'Delayed susceptibility' (category 5) was defined as a steady increase in virus incidence at successively higher apical leaf axes until the highest tested position contained >75% frequency of virus infection. 'Full susceptibility' (category 6) was defined as virus incidence in 100% of apical leaves. The inoculated leaf was tested to discern immunity from early recovery.

Tests for Systemic Recovery

A subset of the host panel exhibiting recovery from inoculation with GFLV strains F13 or GHu [4x(*N. sylvestris* × *N. tomentosiformis*), 4 × (*N.

glutinosa × *N. tabacum*) and 4 × (*N. sylvestris* × *N. otophora*)] was re-inoculated with GFLV-GHu in the fourth leaf position 34 days after the original inoculation. Re-inoculated leaves were tested for GFLV incidence at five dpi by DAS-ELISA. Negative values were interpreted as systemic recovery and positive values were interpreted as a lack of systemic recovery.

Statistics

Statistics were computed on JMP version 10.0 (SAS Institute, Cary, NC). A score of one was assigned for each leaf infected in the first three apical leaf positions, and the sum of these scores among the samples at each leaf position is referred to as virus incidence. Each plant inoculated with a given virus was considered a replicate. Contingency analyses were used to compute Pearson's correlations (r) and contingency tables. Correlation analyses were made for species origin (South America, California, Australia or synthetic), ploidy ($x = 12$ to 48), and virus inoculum (GFLV-F13, GFLV-GHu or ToRSV-AP) with respect to virus incidence at each leaf position. Origin and virus inoculum was considered as categorical variables, ploidy as continuous and virus incidence as ordinal data. A contingency analysis for section was not included due to the limited instances in which multiple species were sampled within a section. Correlation analyses were conducted where synthetic allopolyploids were either included or excluded in the data set.

AUTHORS' CONTRIBUTIONS

JG conceived the research, carried out virus inoculations and tests, analyzed the data and drafted the manuscript. SS and JG performed statistical analyses. RL, SS and MF critically revised the manuscript. RL provided much of the plant material and MF participated in the study design. All authors approved the final manuscript.

ACKNOWLEDGEMENTS

We thank the United States *Nicotiana* Germplasm Collection and researchers at the University of Kentucky for providing *Nicotiana* seeds. We thank Larissa Osterbaan, Drs. John Hart and Christophe Ritzenthaler for helpful discussions, and Dr. Lisa Earle and Ben Orcheski for improving the manuscript. We appreciate the technical assistance provided by Yen-Mei Cheung, Pat Marsella-Herrick and David MacUmber. J. Gottula was supported by a USDA-NIFA-AFRI predoctoral fellowship, M. Fuchs by USDA-NIFA as well as Nolan and Kaplan Funds, S. Saito by California Grape and Tree Fruit League and R. Lewis by N.C. State University.

REFERENCES

1. Clay K, Kover PX: The Red Queen hypothesis and plant/pathogen interactions. *Annu Rev Phytopathol* 1996, 34:29-50.
2. Jackson JA, Tinsley RC: Parasite infectivity to hybridising host species: a link between hybrid resistance and allopolyploid speciation? *Int J Parasitol* 2003, **33:**137-144.
3. Prentis PJ, Wilson JRU, Dormontt EE, Richardson DM, Lowe AJ: Adaptive evolution in invasive species. *Trends Plant Sci* 2008, 13:288-294.
4. Moulia C: Parasitism of plant and animal hybrids: are facts and fates the same? *Ecology* 1999, 80:392-406.
5. Wolinska J, Lively CM, Spaak P: Parasites in hybridizing communities: the Red Queen again? *Trends Parasitol* 2008, 24:121-126.
6. Otto SP, Whitton J: Polyploid incidence and evolution. *Annu Rev Genet* 2000, 34:401-427.
7. Mette MF, Kanno T, Aufsatz W, Jakowitsch J, van der Winden J, Matzke MA, Matzke AJM:Endogenous viral sequences and their potential contribution to heritable virus resistance in plants. *EMBO J* 2002, 21:461-469.
8. Nawaz-ul-Rehman MS, Fauquet CM: Evolution of geminiviruses and their satellites. *FEBS Lett* 2009, 583:1825-1832.
9. Gibbs AJ, Fargette D, García-Arenal F, Gibbs MJ: Time–the emerging dimension of plant virus studies. *J Gen Virol* 2010, 91:13-22.

10. Lewis RS: Nicotiana. In *Wild crop relatives: genomic and breeding resources*. Edited by Kole C. Berlin, Heidelberg: Springer Berlin Heidelberg; 2011:185-208.
11. Chase MW: Molecular systematics, GISH and the origin of hybrid taxa in Nicotiana (Solanaceae). *Ann Bot* 2003, 92:107-127.
12. Kovarik A, Matyášek R, Lim KY, Skalická K, Koukalová B, Knapp S, Chase M, Leitch AR:Concerted evolution of 18–5.8-26S rDNA repeats in Nicotiana allotetraploids. *Biol J Linn Soc* 2004, 82:615-625.
13. Clarkson JJ, Kelly LJ, Leitch AR, Knapp S, Chase MW: Nuclear glutamine synthetase evolution in Nicotiana: phylogenetics and the origins of allotetraploid and homoploid (diploid) hybrids. *Mol Phylogenet Evol* 2010, 55:99-112.
14. Kelly LJ, Leitch AR, Clarkson JJ, Knapp S, Chase MW: Reconstructing the complex evolutionary origin of wild allopolyploid tobaccos (Nicotiana section Suaveolentes). *Evolution* 2013, 67:80-94.
15. Patel D, Power JB, Anthony P, Badakshi F, Heslop-Harrison JSP, Davey MR: Somatic hybrid plants of Nicotiana × sanderae (+) N. debneyi with fungal resistance to Peronospora tabacina. *Ann Bot* 2011, 108:809-819.
16. Renny-Byfield S, Chester M, Kovarik A, Le Comber SC, Grandbastien MA, Deloger M, Nichols RA, Macas J, Novak P, Chase MW, Leitch AR: Next generation sequencing reveals genome downsizing in allotetraploid Nicotiana tabacum, predominantly through the elimination of paternally derived repetitive DNAs. *Mol Biol Evol* 2011, 28:2843-2854.
17. Renny-Byfield S, Kovarik A, Chester M, Nichols RA, Macas J, Novák P, Leitch AR:Independent, rapid and targeted loss of highly repetitive DNA in natural and synthetic allopolyploids of Nicotiana tabacum. *PLoS ONE* 2012, 7:e36963.
18. Clarkson JJ, Lim KY, Kovarik A, Chase MW, Knapp S, Leitch AR: Long-term genome diploidization in allopolyploid Nicotiana section Repandae (Solanaceae). *New Phytol* 2005, 168:241-252.
19. Dawson WO, Hilf ME: Host-range determinants of plant viruses. *Annu Rev Plant Biol* 1992, 43:527-555.

20. Brunt A, Crabtree K, Dallwitz M, Gibbs AJ, Watson L, Zurcher E: Plant viruses online: descriptions and lists from the VIDE database. [http://pvo.bio-mirror.cn/refs.htm*webcite*]
21. Christie SR, Crawford WE: Plant virus range of Nicotiana benthamiana. *Plant Disease Reporter* 1978, 62:20-22.
22. Dijk P, Meer FA, Piron PGM: Accessions of Australian Nicotiana species suitable as indicator hosts in the diagnosis of plant virus diseases. *Eur J Plant Pathol* 1987, 93:73-85.
23. Dijk P, Cuperus C: Reactions on Nicotiana species to potato viruses A, X and Y and tobacco mosaic virus in relation to their taxonomy and geographical origin. *Eur J Plant Pathol* 1989, 95:343-356.
24. Goodin MM, Zaitlin D, Naidu RA, Lommel SA: Nicotiana benthamiana: its history and future as a model for plant–pathogen interactions. *Mol Plant-Microbe Interact* 2008, 21:1015-1026.
25. Yang S, Carter S, Cole A, Cheng NH, Nelson RS: A natural variant of a host RNA-dependent RNA polymerase is associated with increased susceptibility to viruses by Nicotiana benthamiana. *Proc Natl Acad Sci* 2004, 101:6297-6302.
26. Holmes FO: Inheritance of resistance to tobacco-mosaic disease in tobacco. *Phytopathology* 1938, 28:553-561.
27. Cole AB, Király L, Ross K, Schoelz JE: Uncoupling resistance from cell death in the hypersensitive response of Nicotiana species to Cauliflower mosaic virus infection. *Mol Plant-Microbe Interact* 2001, 14:31-41.
28. Taliansky M, Aranda MA, García-Arenal F: Differential invasion by tobamoviruses of Nicotiana megalosiphon following the hypersensitive response. *Phytopathology* 1994, 84:812-815.
29. Fraser R: The genetics of plant-virus interactions: implications for plant breeding. *Euphytica* 1992, 63:175-185.
30. Maule AJ, Caranta C, Boulton MI: Sources of natural resistance to plant viruses: status and prospects. *Mol Plant Pathol* 2007, 8:223-231.
31. Pehu E, Gibson RW, Jones M, Karp A: Studies on the genetic basis of resistance to Potato leaf roll virus, Potato virus Y and Potato virus X in Solanum brevidens using somatic hybrids of Solanum brevidens and Solanum tuberosum. *Plant Sci* 1990, 69:95-101.

32. Lewis RS: Transfer of resistance to potato virus Y (PVY) from Nicotiana africana to Nicotiana tabacum: possible influence of tissue culture on the rate of introgression. *Theor Appl Genet* 2005, 110:678-687.

33. Sanfaçon H, Wellink J, Gall O, Karasev A, Vlugt R, Wetzel T: Secoviridae: a proposed family of plant viruses within the order Picornavirales that combines the families Sequiviridae and Comoviridae, the unassigned genera Cheravirus and Sadwavirus, and the proposed genus Torradovirus. *Arch Virol* 2009, 154:899-907.

34. Andret-Link P, Laporte C, Valat L, Ritzenthaler C, Demangeat G, Vigne E, Laval V, Pfeiffer P, Stussi-Garaud C, Fuchs M: Grapevine fanleaf virus: still a major threat to the grapevine industry. *J Plant Pathol* 2004, 86:183-195.

35. Sanfaçon H, Fuchs M: Tomato ringspot virus. In *Virus and virus-like diseases of pome and stone fruits*. Edited by Hadidi A, Barba M, Candresse T, Jelkmann W. St. Paul, MN: APS Press; 2011:41-48.

36. Raski DJ, Goheen AC, Lider LA, Meredith CP: Strategies against Grapevine fanleaf virus and its nematode vector. *Plant Dis* 1983, 67:335-339.

37. Lamberti F, Golden AM: Redescription of Xiphinema americanum Cobb, 1913 with comments on its morphometric variations. *J Nematol* 1984, 16:204-206.

38. Dias HF, Harrison BD: The relationship between Grapevine fanleaf, Grapevine yellow mosaic and Arabis mosaic viruses. *Ann Appl Biol* 1963, 51:97-105.

39. Jovel J, Walker M, Sanfaçon H: Salicylic acid-dependent restriction of Tomato ringspot virus spread in tobacco is accompanied by a hypersensitive response, local RNA silencing, and moderate systemic resistance. *Mol Plant-Microbe Interact* 2011, 24:706-718.

40. Ratcliff F, Harrison BD, Baulcombe DC: A similarity between viral defense and gene silencing in plants. *Science* 1997, 276:1558-1560.

41. Siddiqui SA, Sarmiento C, Kiisma M, Koivumaki S, Lemmetty A, Truve E, Lehto K: Effects of viral silencing suppressors on Tobacco

ringspot virus infection in two Nicotiana species. *J Gen Virol* 2008, 89:1502-1508.

42. Schwach F, Vaistij FE, Jones L, Baulcombe DC: An RNA-Dependent RNA Polymerase prevents meristem invasion by Potato virus X and is required for the activity but not the production of a systemic silencing signal. *Plant Physiol* 2005, 138:1842-1852.
43. Qu F, Ye X, Hou G, Sato S, Clemente TE, Morris TJ: RDR6 has a broad-spectrum but temperature-dependent antiviral defense role in Nicotiana benthamiana. *J Virol* 2005, 79:15209-15217.
44. Ying X, Dong L, Zhu H, Duan C, Du QS, Lv DQ, Fang Y, García JA, Fang R-X, Guo HS: RNA-Dependent RNA Polymerase 1 from Nicotiana tabacum suppresses RNA silencing and enhances viral infection in Nicotiana benthamiana. *Plant Cell* 2010, 22:1358-1372.
45. Oliver J, Fuchs M: Tolerance and resistance to viruses and their vectors in Vitis sp.: a virologist's perspective of the literature. *Am J Enol Vitic* 2011, 62:438-451.
46. Gottula JW, Lapato D, Cantilina KK, Saito S, Bartlett B, Fuchs M: Genetic variability, evolution and biological effects of Grapevine fanleaf virus satellite RNAs. *Phytopathology* 2013, 103:1180-1187.
47. Vigne E, Gottula J, Schmitt-Keichinger C, Komar V, Ackerer L, Belval L, Rakotomalala L, Lemaire O, Ritzenthaler C, Fuchs M: A strain-specific segment of the RNA-dependent RNA polymerase of Grapevine fanleaf virus determines symptoms in Nicotiana species. *J Gen Virol* 2013, 94:2803-2813.
48. Jovel J, Walker M, Sanfaçon H: Recovery of Nicotiana benthamiana plants from a necrotic response induced by a nepovirus is associated with RNA silencing but not with reduced virus titer. *J Virol* 2007, 81:12285-12297.
49. Pagán I, Alonso-Blanco C, García-Arenal F: The relationship of within-host multiplication and virulence in a plant-virus system. *PLoS ONE* 2007, 2:e786-e786.
50. Ross AF: Localized acquired resistance to plant virus infection in hypersensitive hosts. *Virology* 1961, 14:329-339.
51. Schulze-Lefert P, Panstruga R: A molecular evolutionary concept connecting nonhost resistance, pathogen host range, and pathogen speciation. *Trends Plant Sci* 2011, 16:1360-1386.

52. Gunduz I, Buss GR, Chen P, Tolin SA: Genetic and phenotypic analysis of Soybean mosaic virus resistance in PI 88788 soybean. *Phytopathology* 2004, 94:687-692.
53. Bruening G: Resistance to infection. In *Natural resistance mechanisms of plants to viruses*. Dordrecht: Springer Netherlands; 2006:211-240.
54. Lambrechts L, Fellous S, Koella JC: Coevolutionary interactions between host and parasite genotypes. *Trends Parasitol* 2006, 22:12-16.
55. Fraile A, García-Arenal F: The coevolution of plants and viruses. In *Advances in virus research Volume 76*. Edited by Maramorosch K, Shatkon A, Murphy F. San Diego, CA: Elsevier; 2010:1-32.
56. Fritz RS: Resistance of hybrid plants to herbivores: genes, environment, or both? *Ecology* 1999, 80:382-391.
57. Chen ZJ: Genetic and epigenetic mechanisms for gene expression and phenotypic variation in plant polyploids. *Annu Rev Plant Biol* 2007, 58:377.
58. Rapp RA, Udall JA, Wendel JF: Genomic expression dominance in allopolyploids. *BMC Biol* 2009, 7:18.
59. Scascitelli M, Cognet M, Adams KL: An interspecific plant hybrid shows novel changes in parental splice forms of genes for splicing factors. *Genetics* 2010, 184:975-983.
60. Carr JP, Lewsey MG, Palukaitis P: Signaling in induced resistance. In *Advances in virus research Volume 76*. Edited by Maramorosch K, Shatkon A, Murphy F. San Diego, CA: Elsevier; 2010:57-121.
61. Boatwright JL, Pajerowska-Mukhtar K: Salicylic acid: an old hormone up to new tricks. *Mol Plant Pathol* 2013, 14:623-634.
62. Incarbone M, Dunoyer P: RNA silencing and its suppression: novel insights from in planta analyses. *Trends Plant Sci* 2013, 18:382-392.
63. King KC, Seppälä O, Neiman M: Is more better? Polyploidy and parasite resistance. *Biol Lett* 2012, 8:598-600.
64. Jackson S, Chen ZJ: Genomic and expression plasticity of polyploidy. *Curr Opin Plant Biol* 2010, 13:153-159.
65. Feldman M, Levy AA, Fahima T, Korol A: Genomic asymmetry in allopolyploid plants: wheat as a model. *J Exp Bot* 2012, 63:5045-5059.

66. Soltis PS: Hybridization, speciation and novelty. *J Evol Biol* 2013, 26:291-293.
67. Lou Y, Baldwin IT: Manduca sexta recognition and resistance among allopolyploid Nicotiana host plants. *Proc Natl Acad Sci* 2003, 100:14581-14586.
68. Vallad GE, Goodman RM: Systemic acquired resistance and induced systemic resistance in conventional agriculture. *Crop Sci* 2004, 44:1920-1934.
69. Palloix A, Ayme V, Moury B: Durability of plant major resistance genes to pathogens depends on the genetic background, experimental evidence and consequences for breeding strategies. *New Phytol* 2009, 183:190-199.
70. Floate KD, Whitham TG: The hybrid bridge hypothesis: host shifting via plant hybrid swarms. *Am Nat* 1993, 141:651-662.
71. DeVerna JW, Myers JR, Collins GB: Bypassing prefertilization barriers to hybridization insing in vitro pollination and fertilization. *Theor Appl Genet* 1987, 73:665-671.
72. Ritzenthaler C, Viry M, Pinck M, Margis R, Fuchs M, Pinck L: Complete nucleotide sequence and genetic organization of Grapevine fanleaf nepovirus RNA1. *J Gen Virol* 1991, 72:2357-2365.
73. Serghini MA, Fuchs M, Pinck M, Reinbolt J, Walter B, Pinck L: RNA2 of Grapevine fanleaf virus: sequence analysis and coat protein cistron location. *J Gen Virol* 1990, 71:1433-1441.
74. Huss B, Walter B, Fuchs M: Cross-protection between Arabis mosaic virus and Grapevine fanleaf virus isolates in Chenopodium quinoa. *Ann Appl Biol* 1989, 114:45-60.
75. Bitterlin MW, Gonsalves D: Serological grouping of *Tomato ringspot virus* isolates: implications for diagnosis and cross-protection. *Phytopathology* 1998, 78:278-285.
76. Gottula JW, Lewis R, Saito S, Fuchs M: Virus incidence data from: Allopolyploidy and the evolution of plant virus resistance. *BMC Evol Biol* 2014.

Chapter 7

The Red Queen in the Corn: Agricultural Weeds as Models of Rapid Adaptive Evolution

C C Vigueira[1], K M Olsen[1], and A L Caicedo[2]

[1]Department of Biology, Washington University, St Louis, MO, USA
[2]Biology Department, University of Massachusetts, Amherst, MA, USA

ABSTRACT

Weeds are among the greatest pests of agriculture, causing billions of dollars in crop losses each year. As crop field management practices have changed over the past 12 000 years, weeds have adapted in turn to evade human removal. This evolutionary change can be startlingly rapid, making weeds an appealing system to study evolutionary processes that occur over short periods of time. An understanding of how weeds originate and adapt is needed for successful management; however, relatively little emphasis has been placed on genetically characterizing these systems. Here, we review the current literature on agricultural weed origins and their mechanisms of adaptation. Where possible, we have included examples that have been genetically well characterized. Evidence for three possible, non-mutually exclusive weed origins (from wild species, crop-wild hybrids or directly from crops) is discussed with respect to what is known about the microevolutionary signatures that result from these processes. We also discuss what is known about the genetic basis of adaptive traits in weeds and the range of genetic mechanisms that are responsible. With a better understanding of genetic mechanisms underlying adaptation in weedy species, we can address the more general process of adaptive evolution and what can be expected as we continue to apply selective pressures in agroecosystems around the world.

INTRODUCTION

Domesticated plants have long served as systems for the study of evolution and adaptation (Darwin, 1859). Crop domestication is an evolutionary process brought about by human interactions with plants and manipulation of their natural environment. When wild species are brought into cultivation, they are subjected to a suite of microevolutionary forces. Genetic drift, which randomly fixes alleles, is manifested through population bottlenecks imposed when a subset of individuals from wild populations is cultivated and propagated. Gene flow occurs via hybridization with wild relatives and among crop varieties, resulting in increased diversity across the genome. Strong artificial selection for domestication traits acts to fix alleles in the face of gene flow, and can result in selective sweeps in targeted genomic regions. The signatures of these microevolutionary processes can be sought in crop genomes, as selective sweeps and introgression are predicted to be limited to regions harboring genes involved in key traits, whereas drift and gene flow affect the entire genome (Doebley *et al.*, 2006; Morrell *et al.*, 2011).

As agriculture originated <12 000 years ago (Doebley *et al.*, 2006), all evolutionary changes that take place in the agricultural context must have occurred within this recent time frame. This makes crop domestication an appealing and tractable system for studying the process of rapid evolutionary change. Moreover, because of their tremendous economic importance, many crops benefit from substantial research support, with community investment in genetic and genomic tools for applications in crop improvement. Multiple crop genomes have been sequenced to date, and a handful of these have been re-sequenced in several hundreds of individuals (for example, Lai *et al.*, 2010; Lam *et al.*, 2010). Comparative genomics is now being utilized to reveal similarities and differences in genome evolution during domestication (Morrell *et al.*, 2011).

Although studies of crop domestication have revealed important insights into the genetic mechanisms underlying rapid evolutionary change under human-imposed selection, far less is known about the rapid evolution that can occur in the agricultural environment outside the boundaries of deliberate human selection. When weedy plants occupy this human-mediated environment, they can undergo rapid adaptive evolution to proliferate and escape eradication. Weeds are broadly defined as plants that are 'growing where they are not desired' or 'plants out of place' (Monaco *et al.*, 2002). In this review we focus on weeds that invade and persist in agricultural fields, referred to herein as agricultural weeds. These weeds have evolved to survive in the artificial setting of crop fields, but they are not intentionally selected on by humans and are evolving through unintentional human-mediated selection (Warwick and Stewart, 2005); thus, agricultural weed evolution has occurred under the same forces that guide evolution in natural settings, including genetic drift, gene flow and natural selection.

Agricultural weeds are currently a leading cause of crop losses, resulting in ~10%worldwide reduction in crop productivity (Oerke, 2006) and an estimated $33 billion annual cost to the United States alone (Pimentel *et al.*, 2005). Whereas some of the traits that are adaptive in agricultural weeds are shared with those selectively favored in crops (such as herbicide and disease resistance), other weed-adaptive traits are similar to those favored in wild species (such as seed dispersal and dormancy). Therefore, agricultural weeds compose a unique evolutionary state, neither wild nor domesticated, that has developed in parallel to crop domestication. As study systems, agricultural weeds have many of the same advantages as crops in terms of the recent timing of evolutionary change, but they also offer the additional features of being subject to the forces of natural rather than artificial selection. Beyond their academic interest, understanding how weeds evolve is also clearly important for weed management and protection of the global food supply.

A complete understanding of the occurrence of agricultural weeds requires knowledge of two aspects of weed evolution. First, there is the phylogenetic origin of weed populations: from what wild and/or domesticated ancestors are agricultural weeds derived? Second, there is the origin of weed-adaptive traits *per se*: regardless of weed ancestry, what genetic mechanisms account for the traits that have enabled weeds to successfully invade the agricultural environment? As agricultural weeds are phylogenetically diverse, multiple evolutionary scenarios are likely to account for their appearance. Thousands of species are believed to function as agricultural weeds; about 250 are considered particularly problematic for agriculture worldwide, and these span multiple families and genera (Holm *et al.*, 1977). Much as the current comparative mining of diverse crop genomes is helping us understand the common threads underlying the evolution of traits under domestication, comparative studies of weed evolution can address longstanding questions of how plant groups arise and adapt to novel environments.

There are two broad categories of agricultural weeds: those that are closely related to domesticated species, known as conspecific and congeneric weeds, and those that lack crop relatives. This difference has a key role in shaping weed origins and evolution. In particular, gene flow and introgression from crops is possible for weeds with crop relatives, and these forces influence the mechanisms by which weed-adaptive traits may evolve. For example, herbicide resistance may be acquired through crop-to-weed gene flow for weedy crop relatives, whereas this trait must evolve through *de novo* mutations or from pre-existing (standing) allelic variation in a weed without crop relatives. On the other hand, both classes of weed can potentially exchange genes with wild relatives, which provide an additional source of adaptive variation.

In this review, we survey the state of knowledge regarding the origin of agricultural weeds and their mechanisms of adaptation. We focus primarily on agricultural weed species for which genetic data have contributed to our understanding of their origins and mechanisms of adaptation to the agricultural environment. For more general reviews on agricultural weed evolution, we refer the reader to recent discussions by Ellstrand *et al.* (2010), Stewart *et al.* (2009),Neve *et al.* (2009), and Gressel (2005). Genetic resources are very unevenly distributed among weed taxa; the availability of genetic and genomic information for conspecific or congeneric crop species has often influenced which weed species get studied most. Nonetheless, the emergence of genomic tools that are applicable to non-model species offers nearly universal opportunities to advance the study of agricultural weeds, providing a wealth of study systems that can be used to understand the nature of rapid evolutionary change.

THE ORIGINS OF AGRICULTURAL WEEDS

As the agricultural environment arose <12 000 years ago, plants that invade it represent organisms that have recently occupied new niches. Two potential mechanisms for weed colonization can be envisioned. First, plants that become agricultural weeds may be intrinsically capable of colonizing the agricultural environment. Plants within this group include generalists, with efficient dispersal mechanisms and phenotypically plastic genotypes (described by Baker, 1965), and plants that are specifically pre-adapted to growth and competition for resources in agricultural-like environments. Second, invasion of agricultural settings may be contingent on the process of adaptive evolution. The relative importance of plasticity/pre-adaptation *vs* adaptive evolution in the origins of weeds has been debated for many years in the weed science literature, and it may be that no single mechanism can account for all traits in any particular weed species (Baker, 1965; Clements *et al.*, 2004). Additionally, changes that have occurred in cropping practices over time make it likely that ongoing adaptive evolution is a necessary feature of most successful agricultural weeds (Clements *et al.*, 2004).

De Wet and Harlan (1975) proposed three routes to the origin of agricultural weeds. Figure 1 depicts these scenarios. First, wild colonizers that evolve or already possess traits specific for survival invade agricultural fields. These weeds may experience a demographic bottleneck in association with establishment in the field, resulting in reduced genome-wide neutral genetic diversity. However, gene flow can maintain high levels of genetic diversity if these weeds are in close proximity to populations occurring outside the agricultural setting. Second, hybridization between wild species and domesticated crops can occur, resulting in weeds with an admixed genomic makeup. This diversity can potentially fuel the evolution of adaptive weedy traits; the crop parent may also contribute traits that facilitate a pre-adaptation to crop fields. Weeds in this group may experience a genetic bottleneck if a small number of hybridization events are successful; however, continued hybridization with both wild and crop relatives after establishment could increase genetic diversity. Third, domesticated plants may escape cultivation and evolve into weed populations. Weeds directly descended from crops ('de-domesticates') will likely experience the strongest genetic bottleneck of all, due to the low level of diversity already present in crops compared to wild species. De-domesticated weeds may subsequently hybridize with crops and could hybridize with wild relatives, resulting in increases in genomic diversity.

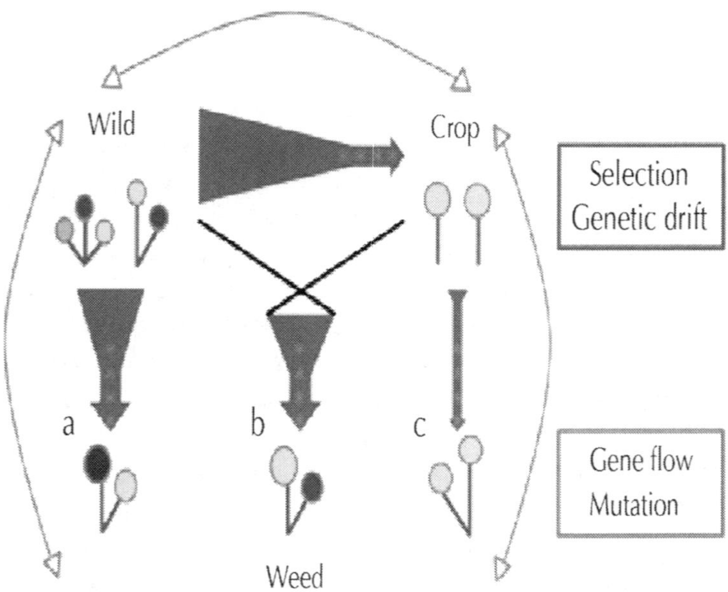

Figure 1: Three alternate origins of agricultural weeds (a) wild colonizers, (b) wild and crop hybridization and (c) crop de-domesticates. Blue arrows represent reductions of diversity due to bottlenecks (genetic drift) and

selection, with relative size indicating expected level of genetic diversity in each scenario. Orange stars represent new mutations and orange arrows represent gene flow through hybridization between groups, both representing increases in genetic diversity. The black X indicates a hybridization event between the crop and wild ancestors leading to the hybrid weed origin. Genes underlying weed adaptive traits come from standing variation, introgression and new mutations.

Although the last two categories of weed origins will result in weeds with close crop relatives, the first category does not eliminate the possibility of a weed having interfertile crop relatives. In practice, it is often difficult to separate out the effects of ancestry and subsequent introgression in weed genomes; thus, unequivocal assignment of one of these routes for the emergence of a given weedy species is not always possible. Additionally, few studies to date have addressed the demographic consequences of weed origins in specific weed species. Below, we discuss the available evidence for each of these routes of weed origins.

Wild to Weed

Colonization of agricultural fields by wild plant species is as old as the practice of agriculture. Whereas selection for traits useful to humans was driving change in early crop species, wild plants were under selection to evade removal from crop fields. Wild plants that become weedy are often generalists, or are pre-adapted to infest agricultural fields. For example, witchweed (*Striga hermonthica*), a parasitic weed of cereal crops, has been shown to withstand a wide range of environmental conditions and can parasitize different hosts (Dawoud and Sauerborn, 1994; Ali *et al.*, 2009). Based on AFLP (amplified fragment length polymorphism) markers, this genetically diverse outcrossing weed does not show genetic differentiation correlated with host specificity, but rather, it is correlated with geography (Welsh and Mohamed, 2011), suggesting that the weed is a generalist that can easily invade new crop hosts. Similarly, no genetic differentiation based on AFLP has been found between populations of broomrape (*Orobanche foetida*) infesting wild legumes and a population parasitizing cultivated vetch, suggesting a generalist ability to undergo host shifts onto cultivated species (Vaz Patto *et al.*, 2008). Blackgrass (*Alopecurus myosuroides*), a non-parasitic weedy grass common in cereal fields worldwide, has high genetic diversity based on amplified fragment length polymorphisms within the populations, and little divergence between populations collected across Europe (Menchari *et al.*, 2007). Local fields are likely connected by high levels of gene flow (Délye *et al.*, 2010), but the lack of differentiation across large ranges suggests that this recently expanded weed did not require adaptive changes to colonize different cereal fields, although adaptation at a few loci cannot be ruled out. Interestingly, the above examples suggest that bottlenecks may not characterize weedy plant populations derived from generalists or from pre-adapted wild plants (Figure 1).

Adaptive evolution has been suggested to have a role in the origin of some wild-to-weed species, although the wild ancestors of weeds tend to be understudied, making conclusive inferences about the mechanisms of weed colonization difficult. One better-studied example of adaptive wild colonization involves the evolution of crop mimicry, the ability to escape human removal by physically resembling the crop in which weeds grow. Phenotypic variation in the progenitors of these weeds could be the source of some mimicry traits early in establishment; however, high levels of specialization have restricted many crop mimics to very specific agricultural settings (Barrett, 1983).

Barnyardgrass (*Echinocloa oryzicola*; Figure 2) is the best characterized crop mimic owing to its worldwide impact on rice yield (Smith, 1988), which makes it one of the costliest of all the agricultural weeds. Despite this impact, there is little genetic information about its evolutionary history, although recent studies have better resolved phylogenetic relationships within *Echinochloa*. Once thought to be conspecific with *E. oryzicola, E. crus-galli* (a separate hexaploid weed species) is now thought to have arisen through the hybridization of *E. oryzicola* (tetraploid) and an unknown diploid species (Aoki and Yamaguchi, 2008). *E. crus-galli* is also a successful weed of several crops but functions as a generalist, whereas *E. oryzicola* has adapted to mimic rice in morphology and phenology. Traits such as large seed size and poor seed dormancy prevent establishment outside of the rice field, suggesting that these traits are adaptive specifically in the agricultural setting (Barrett and Wilson, 1983). In addition, some biotypes possess traits that seem suited to particular farming practices, such as the ability to germinate under water (Smith, 1988). Barnyardgrass has congeneric crops including Indian and Japanese barnyard millet (*E. frumentacea* and *E. esculenta*, respectively) and a conspecific minor crop in China called Mosou barnyard millet, all of which could contribute alleles to the weed. However, ploidy differences between these groups suggest that gene flow from cultivars may not have a role in the weed's evolution.

Figure 2: Some representative agricultural weeds. Clockwise from top left: barnyardgrass in a cultivated rice field (Katie Hyma), waterhemp infesting soybeans (Kate Waselkov), weedy rye infesting cultivated wheat (Jutta Burger), weedy sunflower in a maize field (Nolan Kane), morning glory (Michael Fethe), *Conyza canadensis*(horseweed) invading cultivated maize (Michael Fethe), black hull awned weedy rice in a rice field (Katie Hyma), California wild radish (Norman Ellstrand).

Another example of adaptive evolution enabling wild plants to colonize the agricultural environment is the evolution of herbicide resistance. This strategy is well illustrated by the weedy *Amaranthus* species, collectively known as pigweeds. Several *Amaranthus* crops are widely grown in Central and South America; however, these grain amaranths are genetically distinct from the weedy species, which are believed to have evolved from wild populations (Maughan *et al.*, 2011). A striking example of recent weed invasion is that of waterhemp (*A. tuberculatus*) in corn and soybean fields (Figure 2) in the central United States over the past two decades (Steckel, 2007). Two unresolved hypotheses have been put forth for this phenomenon: (1) changes in weed management such as reduced tilling and reliance on post-emergence herbicides have favored rapid adaptive evolution by selection on standing genetic variation and/or *de novo*mutations (Costea *et al.*, 2005), and (2) hybridization between waterhemp and other weedy *Amaranthus* species has resulted in the acquisition of genes allowing for successful invasion (Trucco *et al.*, 2009). Weedy amaranths will likely continue to be problematic given their ability to rapidly evolve herbicide resistance (discussed below), produce an extraordinary number of seeds and compete for space and nutrients in the field. The evolution of herbicide resistance in wild-to-weed plants has also been reported for *Conyza canadensis* (Zheng *et al.*, 2011;Figure 2), *Lolium rigidum* (Collavo and Sattin, 2012) and barnyardgrass (Hoagland*et al.*, 2004; Talbert and Burgos, 2007), among others.

Hybrid to Weed

Hybridization of crops with wild relatives may result in genetic variants that have traits already adaptive in the agricultural field (those inherited from the crop parent) and traits that favor proliferation and persistence (those inherited from the wild parent), thereby creating a weedy plant. Hybridization can lead to establishment of crop-wild-weed complexes; the possibility of ongoing gene flow in these complexes can increase genome-wide genetic diversity, complicating predictions about the extent of bottlenecks and the major colonization strategies (that is, the roles of preadaptation or adaptive evolution). The expectation is that hybrid weeds may experience a population bottleneck during establishment, but may quickly recover if gene flow with wild and crop relatives persists (Figure 1). However, distinguishing ancient hybridization from ongoing gene flow is challenging, as is differentiating between genetic contributions from crop and wild species when both share ancestral polymorphisms. Examples of weeds with well-documented hybrid origins are described below.

The evolution of weedy sunflowers (*Helianthus annuus*) is well studied in several world regions. Sunflowers are native to North America, where the crop is often grown in proximity to wild *Helianthus* populations. Weedy sunflower strains in European fields most likely originated in seed plots in the US, where nearby populations of wild sunflower contributed pollen to cultivated sunflowers, producing weedy hybrids (Muller *et al.*, 2011). These hybrid seeds were likely exported with cultivated seeds to Europe and established as weeds in the field. Genetic diversity is higher in European populations of weedy sunflower than in the populations of cultivated sunflower, but lower than the populations of US wild sunflower, indicating that the hybrid weeds underwent a small genetic bottleneck during establishment. Transcriptomics has revealed that weedy and cultivated sunflowers from Australia and Israel also show evidence of crop/weed introgression (Lai *et al.*, 2012), indicating that hybridization is not uncommon in this group. Weedy sunflower populations in the United States (Figure 2) have been found to be more closely related to the locally occurring wild populations (Kane and Rieseberg, 2008), suggesting multiple direct wild-to-weed descent events, although ongoing gene flow between local wild and weedy populations cannot be completely ruled out. Regardless, the distinct origins of the United States *vs* old world weeds indicates that there are multiple routes to weediness in the sunflower group. Untangling the relative genomic input from wild and cultivated sunflower within the known hybrid weedy populations may shed light on which genes drive the emergence of weedy traits inherited from each parent in this group.

Cultivated, wild and weedy beets (*Beta vulgaris*) are phylogenetically complicated due to shared ancestry and hybridization. Weedy beets arose from hybridization between cultivated beets grown for seed production and wild inland beets, as indicated by microsatellite and chloroplast markers (Viard *et al.*, 2002; Fénart *et al.*, 2008). Seed beets are presumed to have been pollinated by wild beets and then planted in sugar beet fields. The weeds persist by undergoing self-fertilization (a crop trait), flowering the first year of cultivation (a wild trait), bolting early (a wild trait), and maintaining high levels of seed dormancy (a wild trait), reflecting a mixture of genetic input from wild and cultivated beets (Arnaudet *al.*, 2010). Weedy beet populations do not show evidence of a recent genetic bottleneck, with similar levels of genetic diversity as wild beet populations (Fénartet *al.*, 2008). However, weedy beet populations do show strong spatial structuring, likely due to seed dispersal by gravity and limited pollen flow among weedy populations (Arnaud *et al.*, 2011).

There are several instances where weed origin is unclear, but where crop/weed hybridization currently occurs. These include two weedy sorghum species, johnsongrass (*Sorghum halepense*) and shattercane (*S. bicolor*). Johnsongrass ($2n=40$), a perennial grass that infests several crops and natural areas, can hybridize with cultivated sorghum (*S. bicolor*, $2n=20$). However, it is not clear whether the weed's origin was contingent on this hybridization event. Despite differences in ploidy, genetic introgression from the crop to the weed (through pollen) has been reported in weedy populations (Morrell *et al.*, 2005). This introgression is thought to have contributed to the northward expansion of the weed in North America, providing the genetic variation necessary for a change in the weed's phenology to adapt to colder winters (Warwick *et al.*, 1986). Crop alleles have been reported in weed populations with no recent exposure to cultivated sorghum, indicating that introgressed alleles or weeds themselves have moved across large geographic distances.

Shattercane (*S. bicolor*, $2n=20$) closely resembles the conspecific cultivated sorghum, although it differs from the crop in the ease with which it disperses seeds and its competitive ability, particularly in cornfields (Beckett *et al.*, 1988). Shattercane is an annual plant whose distribution seems limited to regions of the world where sorghum is grown; its origin is unclear (Defelice, 2006), but is often described as a hybrid between cultivated sorghum and various wild relatives (Ejeta and Grenier, 2005). It is clear that shattercane can hybridize with grain sorghum, with introgression moving in the direction of crop to weed, potentially leading to the spread of alleles advantageous in agricultural environments (Ejeta and Grenier, 2005). There are several more weedy sorghum species, but the exact phylogeny has yet to be sorted out (Ejeta and Grenier, 2005), which impedes progress in understanding the origin and adaptation of these weedy species.

Crop to Weed

Though suspected to be a common route to weed emergence, there are few verified examples of crops where volunteer strains have directly evolved into competitive weeds (but see Ellstrand *et al.*, 2010). Weeds evolving directly from crops are evolutionarily intriguing, as they are expected to be more genetically depauperate than other weeds (Figure 1), and sources of adaptive traits should be limited to ancestral standing variation in the crop or new mutations. However, whether such weedy species can arise in the absence of any sort of genetic input from other wild groups is an open question. The best-known examples of crop origins of weedy plants include weedy rye (*Secale cereale*) and weedy rice (*Oryza sativa*), both discussed below. Although there are several other weedy species that have crop relatives, such as the sorghum relatives described above, data showing that the weed arose directly from the crop have not been produced. It is often difficult to distinguish hybrid from crop origins as shared alleles between weeds and crops or wild plants can be due to shared ancestry or introgression. To unravel these complicated phylogenies, genome-wide markers are needed for weed, crop and wild relatives. As these data become increasingly available, it is clear that untangling complex histories of adaptation and gene flow will require further improvement on our theoretical and simulation framework (Strasburg *et al.*, 2012).

Weedy rye (Figure 2), which is particularly problematic in wheat fields, seems to have originated in cultivated rye fields in California (Burger *et al.*, 2006). Volunteer rye has been reported as an agricultural weed in many locations; weedy rye differs from crop volunteers in that it has evolved several distinct traits, including seed shattering, smaller seeds and a delay in flowering time (Burger *et al.*, 2007). Remarkably, these differences have evolved in only about 60 generations, the length of time rye has been cultivated in California. It is possible that flowering time differences evolved first, establishing a reproductive isolating barrier between weed and crop, and allowing for other differences to accumulate quickly in this group. Weedy rye does not show reduced diversity at allozyme and microsatellite markers when compared to its domesticated progenitors, indicating that it did not undergo a detectable genetic bottleneck in the populations studied (Burger *et al.*, 2006). This finding stands in contrast to what might be expected for a recent weed origin from a crop species (Figure 1). An interesting feature of this system is that cultivated rye is itself thought to have originated through domestication of what was originally a weed of wheat and barley fields in Southwest Asia (Zohary, 1971). This implies that rye has evolved from wild to weed to crop to weed, a complex history indeed.

Weedy rice (Figure 2) is an obligate weed of rice worldwide, but populations in different regions of the world are thought to have independent evolutionary origins. Weedy rice in the United States has at least two distinct origins, resulting in two genetically distinct biotypes (Londo and Schaal, 2007; Reagon *et al.*, 2010). Sequence data suggest that the two biotypes have descended from different cultivated Asian rice varieties (Reagon *et al.*, 2010). These de-domestication events likely occurred after the non-shattering seed phenotype was fixed in domesticated Asian rice (discussed below), but before rice cultivation was introduced to the United States (Thurber *et al.*, 2010). Coalescent and Bayesian simulations indicate that the timing of the de-domestication events differs for the two biotypes; however, both exhibit the genetic effects of strong population bottlenecks (Reagon *et al.*, 2010), as expected for weeds descended from crops (Figure 1). Despite the fact that both cultivated and weedy rice are predominantly self-fertilizing, hybridization occasionally occurs, potentially allowing for adaptive traits such as herbicide resistance to be passed to the weed and between the two biotypes (Shivrain *et al.*, 2009; Reagon *et al.*, 2010, 2011). The independent evolution of two weed biotypes in the United States offers a promising opportunity to study the genetics of parallel adaptive evolution.

Beyond the United States, genetic and morphological data for weedy rice populations in some other parts of the world also suggest a de-domesticated origin, including weed strains in China (Cao *et al.*, 2006), Japan (Akasaka *et al.*, 2009) and Bhutan (Ishikawa *et al.*, 2005). There is also evidence that not all weedy rice populations are necessarily descended directly from cultivated ancestors. In regions of the world where the wild progenitor of cultivated rice (*O. rufipogon*) persists, there is potential for wild rice to hybridize with weedy and cultivated rice (Majumder *et al.*, 1997). Genomic data from cultivated, weedy and wild rice in these regions will be useful for determining the origin and level of gene flow between these groups.

THE GENETIC BASIS OF WEED ADAPTATION

From an evolutionary perspective, agricultural weeds capture our attention precisely because of their intrinsic capacity for rapid adaptation. Whether weedy phenotypes represent pre-adaptations or traits acquired during the colonization process, all agricultural weeds must possess traits that permit them to thrive in this recently created environment. Adaptive traits enabling agricultural weeds to survive and thrive vary from species to species, but many are familiar to us as weed-diagnostic traits that favor competitive ability and persistence in the agricultural setting. This set of traits, or the 'agricultural weed syndrome', includes rapid growth, high nutrient use efficiency, seed dormancy, efficient seed dispersal, and crop mimicry and herbicide resistance. Longstanding questions about adaptation particularly applicable to agricultural weeds include the following: (1) what are the relative roles of new mutations, standing variation and introgression in the evolution of weed adaptive traits? (2) At the molecular level, are particular mutation types most often responsible for weedy adaptation? (3) What are the genes that underlie convergent evolution of weediness in different weed species or in different populations of one species? (4) In the arms race with humans, selective pressures have changed through weed history; can we detect the signatures of old *vs* new selection in weed genomes?

Understanding weed adaptation is currently hindered by our incomplete knowledge of the genetic basis of many weedy traits. Even in the few cases where the genetic basis of a weedy trait is well understood, genetic information from both wild and crop relatives is needed to distinguish ancestral (standing) genetic variation from adaptations evolving through *de novo* mutation in weed populations. Given both the economic consequences of weedy traits and the evolutionary appeal of the system, multiple efforts are underway to dissect the genetic basis of weed adaptation. Below, we discuss our current understanding of adaptive trait evolution in agricultural weeds.

Origins of Adaptive Traits

The origin of mutations underlying adaptive traits in organisms has generally been ascribed to one of the three possibilities: standing variation, introgression or new mutation. These are all possible sources of adaptive traits in agricultural weeds regardless of the weed's primary phylogenetic origin. Characterization of the genetic provenance of weedy traits is perhaps most advanced in weedy rice, due to the genomic resources available. Genetic evidence, most of which comes from US weedy rice, points to multiple sources of weed-adaptive traits. We discuss examples of standing variation, new mutation and introgression contributing to weedy rice traits below, followed by examples from other weed species.

Proanthocyanidin pigmentation, resulting in dark reddish-brown color of the pericarp (bran), is one of the most obvious phenotypic characteristics of weedy rice (and the source of the weed's common name, red rice). This trait is also characteristic of wild rice (*O. rufipogon*) and was selected against during crop improvement, leading to most modern domesticated rice cultivars lacking pericarp pigmentation, although some traditional landraces retain the ancestral trait. Loss of function mutations at the *Rc* gene, which encodes a regulatory protein in proanthocyanidin synthesis, are responsible for this transition from red to white pericarps in most domesticated rice (Sweeney *et al.*, 2006). Clustering of weed *Rc*alleles with crop alleles of various red-pigmented varieties, both in US and Asian weeds (Gross *et al.*, 2010; Gu *et al.*, 2011), suggests that weeds have inherited the red pericarp phenotype from standing variation. Pericarp pigmentation likely has a direct role in seed dormancy (Gu *et al.*, 2011), an important adaptive trait for weediness.

The origin of weed alleles is less clear for the shattering phenotype (seed dispersal), a trait that is characteristic of wild and weedy rice and was strongly selected against during domestication. The US, weedy rice strains freely disperse their seeds, whereas their putative Asian domesticated progenitors have reduced shattering (Thurber *et al.*, 2010). Studies of *sh4*, a major gene implicated in loss of shattering during rice domestication, have revealed that weeds carry the derived reduced-function alleles that are fixed in the crop (Thurber *et al.*, 2010). Thus, inheritance of an ancestral functional wild allele is not responsible for shattering in US weedy rice, an observation supported by developmental difference in the shattering phenotype between weeds and wild rice (Thurber *et al.*, 2011). Instead, it appears that the US weeds have re-evolved shattering through *sh4* mutational suppression involving an unknown genetic mechanism. Seed shattering is also characteristic of weedy rice that evolved from a cultivar in Japan, which also shares the *sh4* domestication allele with cultivars (Akasaka *et al.*, 2011). As genes other than *sh4* are involved in the origin of shattering in US and Japanese weeds, novel mutations remain the most likely source for this adaptive trait. However, it is not possible to completely rule out standing variation in the diverse crop ancestor; although cultivated rice does not generally shatter, the range of variation in the crop is large (Thurber *et al.*, 2010), and low-frequency shattering crop variants could exist.

Another likely example of novel mutations giving rise to adaptive weed phenotypes involves the evolution of response to novel temperature cues to avoid untimely germination in Chinese weedy rice, rather than reversion to the ancestral dormancy behavior typical of wild rice (Xia *et al.*, 2011). Specifically, weedy rice collected from northern latitudes was found to germinate at a lower temperature after cold treatment than did weedy rice collected from more southern latitudes, whereas cultivated rice from all sampled regions germinated at the same temperature. Therefore, critical temperature cues for seed germination in Chinese weedy rice appear to correlate with local temperature and latitudinal gradients across the range (Xia *et al.*, 2011). These traits are diverged from rice varieties cultivated in the region, which are thought to be the progenitors of the locally occurring weeds based on simple sequence repeat markers (Cao *et al.*, 2006). As of yet, the gene(s) underlying this potentially adaptive trait are not known.

Introgression has been implicated as a source of adaptive weed traits in numerous cases, but genetic evidence (that is, identifying the introgressed allele that underlies a weedy trait) is often sparse. A subset of US weedy rice has been shown to possess alleles of the growth-related gene, *SD1*, introgressed from local rice cultivars, which are not implicated in the phylogenetic origin of the US weeds. These weeds are shorter, slower growing and earlier flowering than weeds that lack the introgressed allele (Reagon *et al.*, 2011). In this case, the evidence for introgression is clear, but there is no proof yet that this phenotype is adaptive. It is possible that short stature and earlier flowering could be traits that are adaptive for crop mimicry, allowing weeds to escape eradication. Following the introduction of herbicide-resistant rice to US agriculture in the last decade, the emergence of some herbicide-resistant weed strains similarly appears to be linked to crop-weed hybridization (Shivrain *et al.*, 2009).

For weeds that are products of hybridization events, the role of introgression in weed-adaptive phenotypes is often implicitly assumed. Identification of genes underlying adaptive traits in weeds can potentially reveal whether crop or wild alleles are contributing more to weed adaptation, and the extent to which genomic background and novel gene combinations drive weed adaptation. A possible example of this last category occurs in California wild radish (Figure 2), a weedy product of cultivated radish (*Raphanus sativus*) and the wild species jointed charlock (*R. raphanistrum*) (Hegde *et al.*, 2006). California wild radish is notable in that it behaves as both an agricultural weed and occupant of natural areas. It displays higher fitness than both of its progenitors as determined by average fruit weight, rate of survival and reproductive output when grown in common gardens (Ridley and Ellstrand, 2009). These traits were likely acquired from the hybridization event itself; however, as the weeds are obligate outcrossers, heterosis, rather than particular wild or crop alleles, could also provide important contributions to weed fitness (Ridley and Ellstrand, 2009). This has been further supported by the construction of artificial radish hybrids which show elevated fitness in common garden experiments (Campbell *et al.*, 2006).

For plant species that are not crop relatives, the emergence of a weedy life history in crop fields may or may not be due to recent *de novo* adaptive mutations. However, standing variation is believed to be the more likely source of adaptive traits, implying that these wild plants are pre-adapted to agricultural weediness. Unfortunately, there is usually too little information about the status of such agricultural weeds in their native (nonagricultural) environments to allow the determination of which mechanism is most prevalent. Some examples of each of these mechanisms are described below.

Standing variation seems to be involved in the success of *Orobanche* and *Striga*parasitic weeds, as described above. In addition, species that are naturally adapted to disturbed habitats may be pre-adapted to the agricultural environment. This form of pre-adaptation may be the case for pigweed amaranths, including waterhemp (*A. tuberculatus*) and Palmer amaranth (*A. palmeri*). On the other hand, the fact that amaranths have become major agricultural weeds only following the widespread use of herbicides suggests that the invasion of crop fields has occurred at least in part through recent adaptive evolution, in this case, through evolution of resistance against multiple classes of herbicide. Standing variation or pre-adaptation seems to be responsible for herbicide-resistant weedy ryegrass (*Lolium rigidum)* in fields that were never exposed to herbicides (Neve and Powles, 2005). This is also the case for morning glory (*Ipomoea purpurea*) populations that exhibit both tolerance and resistance in fields that were never exposed to glyphosate (Baucom and Mauricio, 2010). Standing variation has been implicated in some wild-to-weed plants that have crop relatives. For example, in a microsatellite study of the US weedy sunflowers, six 'outlier' loci that may have experienced selective sweeps were shared among populations (Kane and Rieseberg, 2008). As each of these populations is thought to have evolved independently from local wild populations, and there is no loss of diversity in weed populations, this suggests that genes necessary for adaptation in agricultural systems exist in the wild (Kane and Rieseberg, 2008).

In contrast to the above examples, the rapid evolution of herbicide resistance in weeds such as barnyardgrass (*E. crus-galli* and *E. oryzicola*) may largely be due to novel, adaptive mutations. Barnyardgrass populations from different world regions have evolved resistance to at least nine different mode of action herbicides via independent mutations (Hoagland *et al.*, 2004; Talbert and Burgos, 2007). The occurrence of these independent mutations as well as the resistance to multiple herbicides within some single populations (Malik *et al.*, 2010) suggests that these traits come from new mutations. Herbicide resistance has likely evolved*de novo* in other weeds due to the fitness cost associated with resistance when herbicide selection is not present. Examples include waterhemp (Duff *et al.*, 2009) and foxtail grass (Darmency *et al.*, 2011).

Mutation Types

Although inferring trait origins is sometimes possible without knowledge of the exact gene underlying the trait, this is not true for assessment of mutation types and genetic convergence. Unfortunately, we do not yet have an extensive list of weed-adaptive genes that we can use to assess evolutionary patterns. Much debate in evolutionary biology has been framed around which types of mutations are more likely to underlie phenotypic and, in particular, adaptive change (for example, Hoekstra and Coyne, 2007; Stern and Orgogozo, 2008). Arguments have been put forth for structural (that is, changing protein sequence) and regulatory mutations (that is, changing gene expression). Parallel discussions on the types of genes involved in crop domestication have suggested that morphological and developmental changes are more likely to involve changes in transcriptional regulation, whereas biochemical and other metabolic changes are more likely to involve loss of function mutations at individual structural genes (Doebley *et al.*, 2006). We suspect the same may be true for genes underlying weed-adaptive traits; however exceptions likely abound, as seen in herbicide resistance mutations discussed below. Additional discussion concerning mutation types has centered on the extent to which convergent evolution in different groups occurs through similar or different genetic mechanisms.

The most studied weed-adaptation genes are those that confer herbicide resistance, as gene identification is facilitated by knowledge of the metabolic targets of many herbicides. Typically, these genes are involved in metabolic pathways that are essential for plant function. Therefore, instead of loss of function mutations, we find mutations that alter the gene (so that the herbicide cannot bind) but maintain function. The mutation types identified so far that result in herbicide resistance in weeds are mostly structural, and include amino acid substitutions, copy number changes and deletions within the gene.

The most common mutations leading to herbicide resistance are amino acid substitutions in the target of the herbicide. One example is the herbicide that targets acetolactate synthase (ALS), also called acetohydroxyacid synthase, an enzyme involved in amino acid biosynthesis. Resistance to ALS herbicides is widespread, with 125 known resistant weedy species to date (Heap, 2012). Twenty-two different amino acid substitutions at seven sites in the *ALS* gene have been shown to cause resistance across multiple species. Similar findings have been found for herbicides that target acetyl-coenzyme A carboxylase, an enzyme needed for lipid metabolism (reviewed in Powles and Yu, 2010).

Palmer amaranth (*A. palmeri*) populations in the US have evolved resistance to several different mode of action herbicides, including glyphosate (Gaines *et al.*, 2010; Heap, 2012). Glyphosate inhibits the enzyme 5-enolpyruvyl-shikimate-3-phosphate synthase (EPSPS). Resistance has been attributed to transposon-mediated gene amplification of EPSPS, which increases EPSPS expression, and presumably provides an excess of the enzyme, which cannot be targeted by the herbicide (Gaines *et al.*, 2010). Between 30 and 50 copies of EPSPS are needed to survive glyphosate rates between 0.5 and 1 kg per ha (Gaines *et al.*, 2011). Resistance first occurred in 2005 in Georgia and has rapidly spread to 11 other states (Heap, 2012). It is not clear whether resistance evolved in the years since glyphosate has been applied or stems from standing variation for gene copy number in the weed population.

Protoxin resistance in waterhemp (*A. tuberculatus*) occurs from a very unusual codon deletion in *PPX2L* (Patzoldt *et al.*, 2006). This gene encodes PPO, an enzyme involved in chlorophyll production. The deletion may have occurred multiple times independently in this group, based on polymorphism in the alleles with the deletion (Lee *et al.*, 2008). Resistance to protoxin is uncommon as most plants contain two gene targets, *PPX1* and *PPX2*, localized to the chloroplast and mitochondria respectively, resulting in two sites of action for this herbicide. *PPX2L*seems to replace the function of the two gene copies, with resistant plants lacking a second gene copy. Susceptible waterhemp plants, on the other hand, have a second *PPX2* copy (Patzoldt *et al.*, 2006; Lee *et al.*, 2008). Therefore, resistance to protoxin may be a result of a loss of a gene copy and a codon deletion in the remaining copy.

Herbicide resistance in weeds is a good example of convergent evolution where both similar and different mutations have been implicated in the convergence of the resistance phenotype among groups or species. For example, *ALS* is a gene repeatedly targeted in the evolution of ALS-herbicide resistance; however multiple mutations at several sites within the gene have been implicated in different weed species, as mentioned above. Conversely, in resistance to triazine herbicides that inhibit photosynthesis, which occurs in 43 different genera worldwide to date (Heap, 2012), the same amino acid substitution is responsible for independent evolution of resistance in all known cases (Gronwald, 1997). Given that there seems to be a common suite of traits that are adaptive for weediness in several species, such as seed shattering, rapid growth and seed dormancy, more opportunities for the study of the genetic basis of weed convergence exist.

WEED EVOLUTIONARY GENETICS: THE PATH AHEAD

The impact of agricultural weeds on humans is immense. Beyond the economic costs caused by crop losses, once weeds are established in agricultural fields, their evolutionary dynamics begin to affect crops. In order to control weeds, humans breed crops and make genetic modifications that allow for the use of herbicides or other methods of control. Weeds then adapt to the changing practices, forcing humans to make further modifications, perpetuating the cycle. Crop–weed interactions offer a view of contemporary rapid evolution, a great example of the arms race between species (Neve *et al.*, 2009). Weeds can also be exploited as crops in regions where nutrients and water limit agricultural potential. For example, the flax mimic, *Camelina sativa* or false flax, has been cultivated as an oilseed crop since the Bronze age with recent renewed interest as a biofuel crop (Hutcheon *et al.*, 2010). Whether false flax was a weed that was brought into cultivation or a crop that became weedy is not clear, but traits such as drought tolerance and low nutrient input make it both a competitive weed and an easy crop (Putnam *et al.*, 1993).

The combination of agricultural importance and unique evolutionary dynamics make agricultural crops a system ripe for the development as a model for studying adaptation and evolution. As genomic tools become more affordable and accessible even for species that are not laboratory model systems, investing in the development of genetic resources for weeds promises to yield great returns at the level of basic science and practical applications. Such calls for genetic and genomic resource development for weed species have already been made (Stewart *et al.*, 2009), and data are starting to accumulate for some weedy species (for example, Anderson *et al.*, 2007; Lee *et al.*, 2009; Parasitic Plant Genome Project at http://ppgp.huck.psu.edu). However, few weedy species have been well characterized genetically. Weedy rice likely is the best genetically characterized weed, but this is due to characterization of its important crop relative. As we move into the genomics era, new tools to study the genetic basis for adaptive evolution such as whole genome sequencing, genome–wide association studies, nested association mapping and many more will allow for a better understanding of the world's most problematic weeds. Ralph Waldo Emerson wrote that a 'weed is a plant whose virtues have not yet been discovered'; we suggest that a weed's greatest virtue is its ability to adapt.

ACKNOWLEDGEMENTS

This study was funded in part by a grant from the US National Science Foundation Plant Genome Research Program (IOS-1032023) to ALC, KMO and Yulin Jia.

REFERENCES

1. Akasaka M, Konishi S, Izawa T, Ushiki J (2011). Histological and genetic characteristics associated with the seed-shattering habit of weedy rice (*Oryza sativa* L.) from Okayama, Japan. *Breed Sci* 61: 168–173.
2. Akasaka M, Ushiki J, Iwata H, Ishikawa R, Ishii T (2009). Genetic relationships and diversity of weedy rice (*Oryza sativa* L.) and cultivated rice varieties in Okayama Prefecture, Japan. *Breed Sci* 59: 401–409.
3. Ali RAMA, El-Hussein AA, Mohamed KI, Babiker AGT (2009). Specificity and genetic relatedness among *Striga hermonthica* strains in Sudan. *Int J Life Sci* 3: 1159–1166.
4. Anderson JV, Horvath DP, Chao WS, Foley ME, Hernandez AG, Thimmapuram J *et al* (2007). Characterization of an EST database for the perennial weed leafy spurge: an important resource for weed biology research. *Weed Sci*55: 193–203.
5. Aoki D, Yamaguchi H (2008). Genetic relationship between *Echinochloa crus-galli* and *Echinochloa oryzicola* accessions inferred from internal transcribed spacer and chloroplast DNA sequences. *Weed Biol Manag* 8: 233–242.
6. Arnaud J-F, Cuguen J, Fénart S (2011). Metapopulation structure and fine-scaled genetic structuring in crop-wild hybrid weed beets. *Heredity* 107: 395–404.
7. Arnaud J-F, Fénart S, Cordellier M, Cuguen J (2010). Populations of weedy crop-wild hybrid beets show contrasting variation in mating system and population genetic structure. *Evol Appl* 3: 305–318.
8. Baker HG (1965). Characteristics and modes of origin of weeds. In: Baker HG, Stebbins GL, (eds). *The Genetics of Colonizing Species*. Academic Press: New York. pp 147–172.

9. Barrett SCH (1983). Crop mimicry in weeds. *Econ Bot* 37: 255–282.
10. Barrett SCH, Wilson BF (1983). Colonizing ability in the *Echinochloa crus-galli*complex (barnyard grass). II. Seed biology. *Can J Bot* 61: 556–562.
11. Baucom RS, Mauricio R (2010). Defence against the herbicide RoundUp predates its widespread use. *Evol Ecol Res* 12: 131–141.
12. Beckett TH, Stoller EW, Wax LM (1988). Interference of four annual weeds in corn (*Zea mays*). *Weed Sci* 36: 764–769.
13. Burger JC, Holt JM, Ellstrand NC (2007). Rapid phenotypic divergence of feral rye from domesticated cereal rye. *Weed Sci* 55: 204–211.
14. Burger JC, Lee S, Ellstrand NC (2006). Origin and genetic structure of feral rye in the western United States. *Mol Ecol* 15: 2527–2539.
15. Campbell LG, Snow AA, Ridley CE (2006). Weed evolution after crop gene introgression: greater survival and fecundity of hybrids in a new environment. *Ecol Lett* 9: 1198–1209.
16. Cao Q, Lu B-R, Xia H, Rong J, Sala F, Spada A *et al* (2006). Genetic diversity and origin of weedy rice (*Oryza sativa* f. *spontanea*) populations found in North-eastern China revealed by simple sequence repeat (SSR) markers. *Ann Bot* 98: 1241–1252.
17. Clements DR, DiTommaso A, Jordan N, Booth BD, Cardina J, Doohan D *et al*(2004). Adaptability of plants invading North American cropland. *Agric Ecosyst Environ* 104: 379–398.
18. Collavo A, Sattin M (2012). Resistance to glyphosate in *Lolium rigidum*selected in Italian perennial crops: bioevaluation, management and molecular bases of target-site resistance. *Weed Res* 52: 16–24.
19. Costea M, Weaver SE, Tardif FJ (2005). The biology of invasive alien plants in Canada. 3. *Amaranthus tuberculatus* (Moq.) Sauer var. *rudis* (Sauer) Costea & Tardif. *Can J Plant Sci* 85: 507–522.
20. Darmency H, Picard JC, Wang T (2011). Fitness costs linked to dinitroaniline resistance mutation in *Setaria*. *Heredity* 107: 80–86.
21. Darwin CR (1859) *On the Origin of Species*. John Murray: London.

22. Dawoud DA, Sauerborn J (1994). Impact of drought stress and temperature on the parasitic weeds *Striga hermonthica* and *Alectra vogelii* in their early growth stages. *Exp Agr* 30: 249–257.
23. De Wet JMJ, Harlan JR (1975). Weeds and domesticates: evolution in the man-made habitat. *Econ Bot* 29: 99–107.
24. Defelice MS (2006). Shattercane, *Sorghum bicolor* (L.) Moench Ssp.*Drummondii* (nees ex steud.) de wet ex davidse—black sheep of the family.*Weed Technol* 20: 1076–1083.
25. Délye C, Clément JAJ, Pernin F, Chauvel B, Le Corre V (2010). High gene flow promotes the genetic homogeneity of arable weed populations at the landscape level. *Basic Appl Ecol* 11: 504–512.
26. Doebley JF, Gaut BS, Smith BD (2006). The molecular genetics of crop domestication. *Cell* 127: 1309–1321.
27. Duff MG, Al-Khatib K, Peterson DE (2009). Relative competitiveness of protoporphyrinogen oxidase-resistant common waterhemp (*Amaranthus rudis*). *Weed Sci* 57: 169–174.
28. Ejeta G, Grenier C (2005). Sorghum and its weedy hybrids. In: Gressel J, (ed.). *Crop Ferality and Volunteerism*. CRC Press: Boca Raton. pp 123–135.
29. Ellstrand NC, Heredia SM, Leak-Garcia JA, Heraty JM, Burger JC, Yao L *et al*(2010). Crops gone wild: evolution of weeds and invasives from domesticated ancestors. *Evol Appl* 3: 494–504.
30. Fénart S, Arnaud J-F, De Cauwer I, Cuguen J (2008). Nuclear and cytoplasmic genetic diversity in weed beet and sugar beet accessions compared to wild relatives: new insights into the genetic relationships within the *Beta vulgaris* complex species. *Theor Appl Genet* 116: 1063–1077.
31. Gaines T, Shaner DL, Ward SM, Leach JE, Preston C, Westra P (2011). Mechanism of resistance of evolved glyphosate-resistant Palmer amaranth (*Amaranthus palmeri*). *J Agric Food Chem* 59: 5886–5889.
32. Gaines T, Zhang W, Wang D, Bukun B, Chisholm ST, Shaner DL *et al* (2010). Gene amplification confers glyphosate resistance in *Amaranthus palmeri*.*Proc Natl Acad Sci USA* 107: 1029–1034.
33. Gressel JB (2005) *Crop Ferality and Volunteerism*. CRC Press: Boca Raton.

34. Gronwald JW (1997). Resistance to PS II inhibitor herbicides. In: De Prado R, Jorrin J, Garcia-Torres L, (eds). *Weed and Crop Resistance to Herbicides*. Kluwer: Dordrecht. pp 53–59.
35. Gross BL, Reagon M, Hsu S-C, Caicedo AL, Jia Y, Olsen KM (2010). Seeing red: the origin of grain pigmentation in US weedy rice. *Mol Ecol* 19: 3380–3393.
36. Gu X-Y, Foley ME, Horvath DP, Anderson JV, Feng J, Zhang L *et al* (2011). Association between seed dormancy and pericarp color is controlled by a pleiotropic gene that regulates abscisic acid and flavonoid synthesis in weedy red rice. *Genetics* 189: 1515–1524.
37. Heap I (2012). International survey of herbicide resistant weeds. http://www.weedscience.org/In.asp (Accessed May 25, 2012).
38. Hegde SG, Nason JD, Clegg JM, Ellstrand NC (2006). The evolution of California's wild radish has resulted in the extinction of its progenitors.*Evolution* 60: 1187–1197.
39. Hoagland RE, Norsworthy JK, Carey F, Talbert RE (2004). Metabolically based resistance to the herbicide propanil in *Echinochloa* species. *Weed Sci*52: 475–486.
40. Hoekstra HE, Coyne JA (2007). The locus of evolution: evo devo and the genetics of adaptation. *Evolution* 61: 995–1016.
41. Holm LG, Plucknett DL, Pancho JV, Herbeger JP (1977) *World's Worst Weeds. Distribution and Biology*. University of Hawaii: Honolulu.
42. Hutcheon C, Ditt RF, Beilstein M, Comai L, Schroeder J, Goldstein E *et al*(2010). Polyploid genome of *Camelina sativa* revealed by isolation of fatty acid synthesis genes. *BMC Plant Biol* 10: 233.
43. Ishikawa R, Toki N, Imai K, Sato YI, Yamagishi H, Shimamoto Y *et al* (2005). Origin of weedy rice grown in Bhutan and the force of genetic diversity.*Genet Resour Crop Evol* 52: 395–403.
44. Kane NC, Rieseberg LH (2008). Genetics and evolution of weedy *Helianthus annuus* populations: adaptation of an agricultural weed. *Mol Ecol* 17: 384–394.
45. Lai J, Li R, Xu X, Jin W, Xu M, Zhao H *et al* (2010). Genome-wide patterns of genetic variation among elite maize inbred lines. *Nat Genet* 42: 1027–1030.

46. Lai Z, Kane NC, Kozik A, Hodgins K, Dlugosch KM, Barker MS *et al* (2012). Genomics of Compositae weeds: EST libraries, microarrays, and evidence of introgression. *Am J Bot* 99: 209–218. |

47. Lam H-M, Xu X, Liu X, Chen W, Yang G, Wong F-L *et al* (2010). Resequencing of 31 wild and cultivated soybean genomes identifies patterns of genetic diversity and selection. *Nat Genet* 42: 1053–1059.

48. Lee RM, Hager AG, Tranel PJ (2008). Prevalence of a novel resistance mechanism to PPO-inhibiting herbicides in waterhemp (*Amaranthus tuberculatus*). *Weed Sci* 56: 371–375.

49. Lee RM, Thimmapuram J, Thinglum KA, Gong G, Hernandez AG, Wright CL *et al* (2009). Sampling the waterhemp (*Amaranthus tuberculatus*) genome using pyrosequencing technology. *Weed Sci* 57: 463–469.

50. Londo JP, Schaal B (2007). Origins and population genetics of weedy red rice in the USA. *Mol Ecol* 16: 4523–4535.

51. Majumder ND, Ram T, Sharma AC (1997). Cytological and morphological variation in hybrid swarms and introgressed population of interspecific hybrids (*Oryza rufipogon* Griff. *Oryza sativa* L.) and its impact on evolution of intermediate types. *Euphytica* 94: 295–302.

52. Malik MS, Burgos NR, Talbert RE (2010). Confirmation and control of propanil-resistant and quinclorac-resistant barnyardgrass (*Echinochloa crus-galli*) in rice. *Weed Technol* 24: 226–233.

53. Maughan P, Smith S, Fairbanks D, Jellen E (2011). Development, characterization, and linkage mapping of single nucleotide polymorphisms in the grain amaranths (*Amaranthus* sp.). *Plant Genome J* 4: 92–101.

54. Menchari Y, Délye C, Le Corre V (2007). Genetic variation and population structure in black-grass (*Alopecurus myosuroides* Huds.), a successful, herbicide-resistant, annual grass weed of winter cereal fields. *Mol Ecol* 16: 3161–3172.

55. Monaco TJ, Weller SC, Ashton FM (2002) *Weed Science: Principles and Practices*. John Wiley and Sons: New York.

56. Morrell PL, Buckler ES, Ross-Ibarra J (2011). Crop genomics: advances and applications. *Nat Rev Genet* 13: 85–96.

57. Morrell PL, Williams-Coplin TD, Lattu AL, Bowers JE, Chandler JM, Paterson AH (2005). Crop-to-weed introgression has impacted allelic composition of johnsongrass populations with and without recent exposure to cultivated sorghum. *Mol Ecol* 14: 2143–2154.

58. Muller M-H, Latreille M, Tollon C (2011). The origin and evolution of a recent agricultural weed: population genetic diversity of weedy populations of sunflower (*Helianthus annuus* L.) in Spain and France. *Evol Appl* 4: 499–514.

59. Neve P, Powles S (2005). High survival frequencies at low herbicide use rates in populations of *Lolium rigidum* result in rapid evolution of herbicide resistance. *Heredity* 95: 485–492.

60. Neve P, Vila-Aiub M, Roux F (2009). Evolutionary-thinking in agricultural weed management. *New Phytol* 184: 783–793.

61. Oerke EC (2006). Crop losses to pests. *J Agric Sci* 144: 31.

62. Patzoldt WL, Hager AG, McCormick JS, Tranel PJ (2006). A codon deletion confers resistance to herbicides inhibiting protoporphyrinogen oxidase. *Proc Natl Acad Sci USA* 103: 12329–12334.

63. Pimentel D, Zuniga R, Morrison D (2005). Update on the environmental and economic costs associated with alien-invasive species in the United States.*Ecol Econ* 52: 273–288.

64. Powles SB, Yu Q (2010). Evolution in action: plants resistant to herbicides.*Annu Rev Plant Biol* 61: 317–347.

65. Putnam DH, Budin JT, Field LA, Breene WM (1993). *Camelina*: a promising low-input oilseed. In: Janick J, Simon JE, (eds). *New Crops*. Wiley: New York. pp 314–322.

66. Reagon M, Thurber CS, Gross BL, Olsen KM, Jia Y, Caicedo AL (2010). Genomic patterns of nucleotide diversity in divergent populations of US weedy rice. *BMC Evol Biol* 10: 180–196.

67. Reagon M, Thurber CS, Olsen KM, Jia Y, Caicedo AL (2011). The long and the short of it: SD1 polymorphism and the evolution of growth trait divergence in US weedy rice. *Mol Ecol* 20: 3743–3756.

68. Ridley CE, Ellstrand NC (2009). Evolution of enhanced reproduction in the hybrid-derived invasive, California wild radish (*Raphanus sativus*). *Biol Invasions* 11: 2251–2264.

69. Shivrain VK, Burgos NR, Sales MA, Mauromoustakos A, Gealy DR, Smith KL *et al* (2009). Factors affecting the outcrossing rate between Clearfield™ rice and red rice (*Oryza sativa*). *Weed Sci* 57: 394–403.

70. Smith RJ (1988). Weed thresholds in southern US rice, *Oryza sativa*. *Weed Technol* 2: 232–241.

71. Steckel LE (2007). The dioecious *Amaranthus* spp.: here to stay. *Weed Technol* 21: 567–570.

72. Stern DL, Orgogozo V (2008). The loci of evolution: how predictable is genetic evolution? *Evolution* 62: 2155–2177.

73. Stewart CN, Tranel PJ, Horvath DP, Anderson JV, Rieseberg LH, Westwood JH *et al* (2009). Evolution of weediness and invasiveness: charting the course for weed genomics. *Weed Sci* 57: 451–462.

74. Strasburg JL, Sherman NA, Wright KM, Moyle LC, Willis JH, Rieseberg LH (2012). What can patterns of differentiation across plant genomes tell us about adaptation and speciation? *Philos Trans R Soc Lond B Biol Sci* 367: 364–373.

75. Sweeney MT, Thomson MJ, Pfeil BE, McCouch S (2006). Caught red-handed: Rc encodes a basic helix-loop-helix protein conditioning red pericarp in rice. *Plant Cell* 18: 283–294.

76. Talbert RE, Burgos NR (2007). History and management of herbicide-resistant barnyardgrass (*Echinochloa crus-galli*) in Arkansas rice. *Weed Technol* 21: 324–331.

77. Thurber CS, Hepler PK, Caicedo AL (2011). Timing is everything: early degradation of abscission layer is associated with increased seed shattering in US weedy rice. *BMC Plant Biol* 11: 14.

78. Thurber CS, Reagon M, Gross BL, Olsen KM, Jia Y, Caicedo AL (2010). Molecular evolution of shattering loci in US weedy rice. *Mol Ecol* 19: 3271–3284.

79. Trucco F, Tatum T, Rayburn AL, Tranel PJ (2009). Out of the swamp: unidirectional hybridization with weedy species may explain the prevalence of *Amaranthus tuberculatus* as a weed. *New Phytol* 184: 819–827.

80. Vaz Patto MC, Diaz-Ruiz R, Satovic Z, Roman B, Pujadas-Salva AJ, Rubiales DR (2008). Genetic diversity of Moroccan populations of *Orobanche foetida*: evolving from parasitising wild hosts to crop plants. *Weed Res* 48: 179–186.

81. Viard F, Bernard J, Desplanque B (2002). Populations of weedy crop-wild hybrid beets show contrasting variation in mating system and population genetic structure. *Theor Appl Genet* 104: 688–697.

82. Warwick SI, Phillips D, Andrews C (1986). Rhizome depth: the critical factor in winter survival of *Sorghum halepense* (L.) Pers. (johnson grass). *Weed Res* 26: 381–388.

83. Warwick SI, Stewart CN (2005). Crops come from wild plants—how domestication, transgenes, and linkage together shape ferality. In: Gressel JB, (ed.). *Crop Ferality and Volunteerism*. CRC Press: Boca Raton. pp 9–30.

84. Welsh AB, Mohamed KI (2011). Genetic diversity of *Striga hermonthica*populations in Ethiopia: evaluating the role of geography and host specificity in shaping population structure. *Int J Plant Sci* 172: 773–782.

85. Xia H-B, Xia H, Ellstrand NC, Yang C, Lu B-R (2011). Rapid evolutionary divergence and ecotypic diversification of germination behavior in weedy rice populations. *New Phytol* 191: 1119–1127.

86. Zheng D, Kruger GR, Singh S, Davis VM, Tranel PJ, Weller SC *et al* (2011). Cross-resistance of horseweed (*Conyza canadensis*) populations with three different ALS mutations. *Pest Manag Sci* 67: 1486–1492.

87. Zohary D (1971). Origin of south-west Asiatic cereals: wheats, barley, oats, and rye. In: Davis P, Harper P, Hedge I, (eds). *Plant Life in South-West Asia*. Royal Botanic Garden: Edinburgh. pp 235–263.

Chapter 8

Polyploidy and its Effect on Evolutionary Success: Old Questions Revisited with New Tools

A Madlung[1]

[1]Department of Biology, University of Puget Sound, Tacoma, WA, USA

ABSTRACT

Polyploidy, the condition of possessing more than two complete genomes in a cell, has intrigued biologists for almost a century. Polyploidy is found in many plants and some animal species and today we know that polyploidy has had a role in the evolution of all angiosperms. Despite its widespread occurrence, the direct effect of polyploidy on evolutionary success of a species is still largely unknown. Over the years many attractive hypotheses have been proposed in an attempt to assign functionality to the increased content of a duplicated genome. Among these hypotheses are the proposal that genome doubling confers distinct advantages to a polyploid and that these advantages allow polyploids to thrive in environments that pose challenges to the polyploid's diploid progenitors. This article revisits these long-standing questions and explores how the integration of recent genomic developments with ecological, physiological and evolutionary perspectives has contributed to addressing unresolved problems about the role of polyploidy. Although unsatisfactory, the current conclusion has to be that despite significant progress, there still isn't enough information to unequivocally answer many unresolved questions about cause and effect of polyploidy on evolutionary success of a species. There is, however, reason to believe that the increasingly integrative approaches discussed here should allow us in the future to make more direct connections between the effects of polyploidy on the genome and the responses this condition elicits from the organism living in its natural environment.

INTRODUCTION

Polyploids are organisms whose genomes consist of more than two complete sets of chromosomes. Stebbins distinguished three major types of polyploids: autopolyploids, allopolyploids and segmental allopolyploids (Stebbins, 1947). In autopolyploids all genomes are identical or very similar and arise via genome duplication within the same species (Stebbins, 1947; Lewis, 1980). Allopolyploids, by contrast, contain two or more distinct genomes, and can arise via hybridization of two different species concomitant with genome doubling (Stebbins, 1947;Grant, 1975). Segmental allopolyploids carry more than two partially differentiated genomes, which can lead to the formation of both bivalents and multivalents during chromosome pairing (Stebbins, 1947; Levin, 2002).

Major routes of polyploid formation are via gametic non-reduction and, to a lesser degree, somatic doubling (Lewis, 1980; Ramsey and Schemske, 1998; Levin, 2002;Coyne and Orr, 2004). In gametic non-reduction, fusion of two gametes, of which at least one contains a non-reduced, full somatic complement of chromosomes, can lead to polyploidy. Somatic doubling may occur in zygotic, embryonic or sporophytic tissue. Spontaneous genome duplication in those tissues can thus also produce viable polyploid offspring via gamete formation in the duplicated sectors (Ramsey and Schemske, 1998).

Polyploidization in general is now believed to have occurred at least once during the evolutionary history of all angiosperms (Jiao *et al.*, 2011) and it has been suggested as a major driving force of plant evolution (Blanc and Wolfe, 2004;Chen, 2007; Soltis and Soltis, 2009). Indeed, polyploidization has been described as a process leading to instantaneous speciation (Mayr, 1963) because a single-generation event such as the hybridization between two species with subsequent somatic doubling, or the fusion of unreduced gametes is enough to quickly establish barriers that prevent gene flow between the new polyploid and the old progenitor species, and can lead to reproductive isolation (Ramsey and Schemske, 1998).

Although auto- and allopolyploidy share the property of duplicated genomes, the difference in their compositions has important consequences. In autopolyploids, chromosomes generally pair as multivalents during meiosis, while in allopolyploids bivalent pairing between chromosomes of the same original genome is prevalent (Stebbins, 1947), resulting largely in the maintenance of two separate genomes. However, the more closely genomes in the allopolyploid are related, the more likely it is for homoeologs (the chromosomes duplicated by allopolyploidy) to pair (Levin, 2002), resulting in chromosomal exchanges between the two genomes.

Despite the interest in polyploidy there are many long-standing, yet still unanswered questions about the significance of genome doubling, especially in the areas of ecology, physiology and evolution (Soltis *et al.*, 2010). In this review, I revisit two of these questions and discuss how recent genomic developments have furthered the discussion on these questions. I aim to pay special attention to areas where complementary approaches of different disciplines are starting to be synergistic and where continued integration of genomic with ecological and organismal information would help in the future to resolve unanswered problems.

IS POLYPLOIDY ADVANTAGEOUS FOR EVOLUTIONARY SUCCESS?

There has been a long-standing debate over the evolutionary implications of genome doubling (Winge, 1917; Müntzing, 1936;

Ohno, 1970; Comai, 2005; Otto, 2007; Soltis and Soltis, 2009), which span the spectrum from polyploidy being an evolutionary dead end (Stebbins, 1950; Arrigo and Barker, 2012) to polyploidy being a mayor player with a significant role in evolution (Ohno, 1970; Chen, 2010;Mayfield *et al.*, 2011).

Three major advantages are often cited that should give polyploids an edge over their diploid parents. First, the increased number of alleles of a given gene in a polyploid should allow the masking of deleterious recessive mutations and thus insure against the loss of fitness (Gu *et al.*, 2003). The second proposed advantage of allopolyploids and heterozygous autopolyploids is that heterosis allows offspring to display transgressive performance compared with its progenitor species (Birchler *et al.*, 2010). In contrast to diploid hybrids where hybrid vigor decays over subsequent generations due to homologous recombination, heterosis is stable in allopolyploids due to the predominant disomic pairing of identical homologous chromosomes (Ohno, 1970; Comai, 2005). The third major advantage of polyploids stems from the possibility that duplicated gene copies can evolve to assume new or slightly varied functions (neofunctionalization or subfunctionalization), potentially allowing for ecological niche expansion or increased flexibility in the organism's responsiveness to environmental change (Adams and Wendel, 2005; Moore and Purugganan, 2005; Lynch, 2007).

Among the disadvantages that could lead to less vigor and a reduced adaptive capacity in polyploids are the increased number of chromosomes, and the greater complexity of their pairing and segregation interactions that can cause abnormalities (including aneuploidy) during meiosis and mitosis (Comai, 2005). Additionally, the cell architecture in polyploids is altered because of generally increased cell size in polyploids, which alters the surface to volume ratio (Speckman *et al.*, 1965; Rhoades and Dempsey, 1966; Melaragno *et al.*, 1993;Levin, 2002). Of particular concern to newly formed outcrossing polyploids is the minority cytotype disadvantage. As pollination is more likely to occur with haploid pollen from neighboring diploid progenitor individuals than with diploid pollen from another polyploid, the major proportion of the offspring is likely to be triploid. Triploid individuals are then less likely to be viable, or if they are viable, they are less likely to be fertile (Levin, 1975). Finally, changes in polyploids that can be either advantageous or detrimental are the altered transcriptome, genomic architecture and epigenetic landscape, which can lead to gene silencing or activation, as well as DNA loss and epigenetic changes (Wang *et al.*, 2004, 2006;Hegarty *et al.*, 2006; Gaeta *et al.*, 2007, 2009; Chaudhary *et al.*, 2009; Pumphrey*et al.*, 2009; Rapp *et al.*, 2009; Wright *et al.*, 2009; Buggs *et al.*, 2010, 2011;Hegarty *et al.*, 2011; Mayfield *et al.*, 2011; Xiong *et al.*, 2011; Matsushita *et al.*, 2012). In the new polyploid such changes can lead to reduced, additive or transgressive states. Therefore, polyploidization produces offspring that has the potential to be distinctly different from both of its parents.

Despite the shortage of convincing evidence linking genomic change to physiologically based enhancement of traits and proven evolutionary success (as opposed to just the theoretical promise of evolutionary success), many of the recent articles cited above have concluded that polyploidy has a definitive role in evolution, and questioned or ignored Stebbins's notion that polyploidy might be an evolutionary dead end (Stebbins, 1950). Recently, investigators using modern phylogenetic approaches have revived the debate over the question to what degree polyploidy has a creative role in evolution (Wood *et al.*, 2009; Mayrose *et al.*, 2011; Arrigo and Barker, 2012). Wood *et al.* (2009) reported that while 15% of all speciation events in angiosperms coincided with an increase in ploidy, subsequent faster speciation rates—as might be expected if the duplicated genome indeed allowed for the new polyploid species to populate new ecological niches or have a broader response to environmental change—did not seem to be the norm for these species (Wood *et al.*, 2009). Using a comprehensive phylogenetic data set and estimating rates of diversification in newly formed polyploids, Mayrose *et al.* (2011) determined that polyploids were more likely to be found at the branch tips of phylogenetic trees, and concluded that polyploidy was often followed by extinction as opposed to diversification. Their analyses further showed that speciation rates of polyploid plants were on average lower than those in diploid plants (Mayrose *et al.*, 2011), suggesting that polyploidy is actually more likely to be detrimental than advantageous for evolutionary success. From these two studies one might infer (a) that polyploidy is a condition that occurs frequently and accidentally but does not generally lead to further speciation events after polyploidization, (b) that evolutionary success of neopolyploids is possible but not the norm and (c) that extinction of recently formed species occurs more frequently in polyploids than in diploid relatives

(Arrigo and Barker, 2012). On the other hand, Mayrose *et al.* (2011) also suggested that increased genomic diversity in neopolyploids could be the key to success in those new polyploids that did persevere during the establishment of the new species (Mayrose *et al.*, 2011).

In considering the implications of the two phylogenetic studies described in the previous paragraph (Wood *et al.*, 2009; Mayrose *et al.*, 2011), it is important to note that neither study specifically distinguished between auto- and allopolyploidy. These two conditions can be difficult to distinguish phylogenetically but the differences between allopolyploidy versus auto-polyploidy can have significant and different consequences on the organism (Grant, 1971; Levin, 2002), and thus likely its potential for evolutionary success. If auto- and allopolyploids indeed have different rates of success in evolution, it is important to study their responses to polyploidy separately if the question is whether genome doubling is advantageous or disadvantageous. While it is not trivial to distinguish between the effects of hybridization and genome doubling in allopolyploidization events, distinct differences between genome doubling and hybrid effects were described in several recent studies in *Arabidopsis, Senecio* (ragwort) and*Gossypium* (cotton). In *Arabidopsis*, Wang *et al.* (2006) used microarrays to measure transcriptome changes in neoallopolyploids compared with their autotetraploid parents and found hundreds of genes to be differentially regulated between parents and offspring. The same study, using autopolyploids that had gone through whole-genome duplication but not through hybridization, displayed considerably fewer changes between tetraploids and diploid progenitors. This led the authors to suggest that hybridization, not genome duplication, was the driving force in transcriptomic change in allopolyploids (Wang *et al.*, 2006; Chen, 2007). Two subsequent studies that analyzed transcriptomic changes in *Arabidopsis*autopolyploids only partially supported the view that hybridization resulted in greater genomic change than genome doubling. Pignatta *et al.* (2010) used three independently produced sets of autopolyploids and diploids that arose from differently affected sectors of colchicine-treated parent *Arabidopsis* (ecotype Landsberg *erecta*, L*er*) individuals. While they found no genes that were consistently differentially regulated in response to genome doubling in their microarray analysis, they did observe significant differences between transgenic diploids and tetraploids carrying tissue-specific green fluorescent protein genes activated by endogenous enhancer elements ('enhancer traps'). Differential expression of these enhancers between diploids and tetraploids suggested that perhaps a less complex and more ploidy-sensitive regulation of enhancer traps compared with native genes was responsible for this difference (Pignatta *et al.*, 2010). Yu *et al.* (2010) reported that the magnitude of transcriptomic changes in response to genome doubling in diploid and autotetraploid *Arabidopsis* was dependent on the ecotype used. These authors reported several hundred differentially expressed genes in seedlings or leaves in the ecotype Columbia, but only 9 and 22 genes that were differentially expressed in seedlings and leaves, respectively, in the ecotype L*er* (Yu *et al.*, 2010), the same ecotype that Wang*et al.* (2006) had used in their study of transcriptomic changes in auto- and allopolyploids.

Using the *Senecio* system, Hegarty *et al.* (2005, 2006) studied the separate effects of hybridization and genome duplication by first comparing the synthetic triploid hybrid *Senecio x baxteri* with its two progentior species, the diploid *S. squalidus* and the tetraploid *S. vulgaris*. Subsequently, genome doubling was induced in the triploid using colchicine (Hegarty *et al.*, 2005, 2006) to generate hexaploids equivalent to the naturally occurring *S. cambrensis*. Microarray-based transcriptome analysis in this system suggested that both hybridization and genome doubling had profound effects on the transcriptome but that these effects were distinct from each other (Hegarty *et al.*, 2005, 2006). Furthermore, the authors reported that genome doubling reduced the number of differentially regulated genes in the hybrid compared with the number of differentially regulated genes in the allohexaploid (when either one was compared with their progenitor species), and suggested that genome doubling, therefore, has a 'calming effect' on the hybridization-induced transcriptome shock (Hegarty *et al.*, 2006).

To distinguish between the effects of hybridization and allopolyploidization, Flagel*et al.* (2008) compared gene expression changes between a synthetically produced F1 hybrid and those in a natural allopolyploid of the same original parentage in cotton (*Gossypium*). These authors reported that around 24% of genes with a significantly differential regulation in the F1 hybrid were also found to have the same expression bias in the natural allopolyploid when compared with gene expression in the parents, suggesting that these genes were altered due to genome merger (Flagel *et al.*, 2008). The remaining 76% of expression changes that were observed in the natural allopolyploid (compared with its progenitor species) could reflect accumulated mutations and sub- or neofunctionalized duplicate genes that had assumed altered expression patterns over evolutionary time in the polyploid (Flagel *et al.*, 2008). Although the experimental design of this study did not allow the authors to make direct inferences of the effect of genome doubling on changes in gene expression, they were able to conclude that genome merger had a significant but likely only partial role in altering expression patterns between diploids and hybrids or allopolyploids (Flagel *et al.*, 2008).

From the discussion of the studies cited above (Hegarty *et al.*, 2005, 2006; Wang*et al.*, 2006; Flagel *et al.*, 2008; Pignatta *et al.*, 2010; Yu *et al.*, 2010) it appears that different species, and even different ecotypes of the same species, can vary significantly in their responses to hybridization, genome doubling or the combination of the two. If we assume that the differential expression of the genome in all polyploids compared with their parents has a role in the evolutionary trajectory of the neopolyploids, then it is tempting to speculate that the magnitude of transcriptomic change in the neopolyploid (or the specific change of transcriptional activity in some key genes) can have a deciding role over which polyploids will persevere in evolution, and for which polyploids genome duplication leads to a dead end. A similarly important aspect that warrants more investigation is the question to what degree the genomic changes that occur during polyploidization lead to increased variation in the sibling offspring. Whether or not polyploidy induces greater variation between siblings than is found in diploid populations is still an open question (Lee and Chen, 2001; Gaeta *et al.*, 2007;Madlung *et al.*, 2012). It is important to realize that despite the increased variation—and thus evolutionary potential—described in many of the genomic studies discussed in this article, survival or vigor of neopolyploids in the*Arabidopsis*, *Brassica* and corn study systems have often been low in synthetic populations (Comai *et al.*, 2000; Madlung *et al.*, 2005; Riddle *et al.*, 2006; Gaeta*et al.*, 2007; Matsushita *et al.*, 2012), suggesting either a severe bottleneck for the establishment of vigorous polyploids (Comai, 2005), or the beginning of extinction. Likewise, and very importantly, we still cannot infer from the integration of the molecular and phylogenetic studies discussed earlier whether there is a difference in polyploid advantage or disadvantage for autopolyploids compared with allopolyploids. Molecular analysis of more species may help determine if significant transcriptome changes in response to genome duplication are the rule or the exception as seen in the example of ecotype variation in*Arabidopsis* (Wang *et al.*, 2006; Yu *et al.*, 2010). And finally, to determine if polyploidy-induced genomic changes are adaptive, we also need more integrative studies assessing the physiological and ecological consequences of the genomic changes.

Despite the relative rarity of radiation associated with polyploid speciation discussed earlier (Wood *et al.*, 2009; Mayrose *et al.*, 2011; Arrigo and Barker, 2012), it is important to recognize the evolutionary potential of polyploidy. For that purpose, I will briefly discuss two individual examples of older evolutionarily successful polyploids, which not only survived polyploidization, but for which genome doubling were followed by additional speciation events. Since its split from a common ancestor, *Gossypium* has evolved into three distinct lineages throughout the world: the New World clade (or D-genome); the African–Asian clade separated into the A-, B-, E- and F-genomes; and the Australian clade with the C-, G- and K-genomes (Wendel and Cronn, 2003). Ancestral lines of*Gossypium* are diploid, but an allopolyploidization event took place probably between 1–2 mya after long-distance dispersal of the A-genome to the New World, where it hybridized with the D-genome (Wendel and Cronn, 2003). This new allopolyploid has since radiated into five separate, morphologically diverse species: *G. mustelinum, G. darwinii, G. barbadense, G. tomentosum*, and the most important global supplier of agriculturally used cotton fiber *G. hirsutm*(Fryxell, 1979; Wendel and Cronn, 2003). A second example of an older allopolyploid that has successfully diversified is the genus *Glycine* (soybean).*Glycine* is a complex genus that underwent a whole-genome duplication an estimated 15 mya (Schlueter *et al.*, 2004) and multiple fairly recent allopolyploidization events after its colonization of the Australian continent, for example, with the formation of *G. tabacina* about 30 K years ago (Doyle *et al.*, 1999). In another polyploid species in the genus *Glycine, G. tomentella*, multiple separate 'races' are recognized that have not yet taxonomically arisen to species status, but that display enough molecular phylogenetic divergence from one another to be grouped into separate taxa (Doyle *et al.*, 2004). These two examples from cotton and soybean suggest that diversity and speciation may indeed be found in nature in relatively recent polyploids, exemplifying how molecular diversity induced by polyploidization might have been the driving force in these speciation events. These two case studies are consistent with the possibility that polyploidization in some species can lead to subsequent diversification, when genome doubling in many other species was followed by extinction (Arrigo and Barker, 2012). Nonetheless, it is also important in this context to note that the coincidence of polyploidy and subsequent speciation does not preclude other mechanisms that may have had a role in the evolutionary success of cotton and soybean, such as gene flow between populations, recurrent polyploid origins and hybridization or introgression between different species after polyploidization (Coyne and Orr, 2004).

DO POLYPLOIDS HAVE A GREATER POTENTIAL FOR ADAPTATION TO VARIABLE OR STRESSFUL ENVIRONMENTS?

In the context of evolutionary success of polyploidy the question whether or not polyploids can adapt better to stressful environments was raised early on and has been studied extensively, and discussed at times controversially, in the older literature (Hagerup, 1932; Müntzing, 1936; Gustavsson, 1946; Löve and Löve, 1949; Stebbins, 1950; Löve, 1953; Grant, 1971). Polyploids have been reported to be more frequent in extreme environments, including the subarctic regions, high elevations and xeric environments (Löve and Löve, 1949; Löve, 1953; Hanelt, 1966; Grant, 1971). Given the hypothesis that polyploids can thrive relative to their diploid progenitors, polyploids have been studied for their morphological, physiological and developmental differences from diploids to find correlative evidence that might explain observations of higher stress endurance. One example of morphological differences between diploids and polyploids are the larger cell sizes in polyploids (Melaragno *et al.*, 1993) including those of the stomata (Speckman *et al.*, 1965; Masterson, 1994; Hodgson *et al.*, 2010). Changes in stomatal pore size in response to polyploidy could suggest an effect of genome duplication on the water relations of the plant. A lower density of stomata in polyploids compared with diploids as found in *Betula payrifera* (birch) (Li *et al.*, 1996) could on the other hand reduce or offset any differences in the overall gas exchange rates. Additionally, morphological variation in leaf thickness, pubescence, cuticular thickness or -composition, as well as physiological parameters affecting the plant's water potential can also have a role in water relations and gas exchange rates (Johnson, 1975; Li *et al.*, 1996).

Integrative and experimental approaches to questions regarding the adaptive role of polyploidy in responses to the environment are still rare. Maherali *et al.* (2009)compared natural diploid, tetraploid and colchicine-induced neotetraploids of fireweed (*Chamerion angustifolium*) to study the effect of genome doubling on water relations. The authors reported larger stomata, increased stem and vessel diameter, and decreased specific hydraulic conductivity in both types of tetraploids over the diploids, but noted that the established tetraploids showed significantly greater drought tolerance than the similarly responding diploid and neotetraploid, suggesting that drought tolerance in fireweed likely evolved after and independently of genome duplication (Maherali *et al.*, 2009). Ramsey (2011)used transplant experiments in the field to test survivorship of seedlings raised in the greenhouse and transplanted into a dry dune habitat. He reported a fivefold greater survivorship of seedlings raised from field collected seeds of hexaploids over tetraploids, and a 70% greater survivorship of seedlings raised from seeds of spontaneous neohexaploids compared with tetraploids (Ramsey, 2011). Ramsey concluded from these experiments that polyploidy per se can have a role in adaptation to a new environment (Ramsey, 2011).

In another recent study with an integrative experimental design examining the ecophysiological implications of an altered morphology in polyploids, Li *et al.*(2012) examined the underlying mechanisms of cell size variation in diploid and tetraploid *Arabidopsis thaliana* with molecular tools. Using quantitative PCR, these authors queried expression differences in 34 cell cycle regulating genes and reported significant differences in the expression of three inhibitors of cyclin dependent kinases (*ICK1, 2, 5*) between diploids and tetraploids. These results suggest a mechanism by which altered gene expression levels of select genes could have direct results on the morphology and ecology of the plant. The results by Li *et al.* (2012) set the stage for follow-up experiments that could, for example, test if experimentally altered gene expression of *ICK1, 2* and *5* has direct or proportional consequences on stoma size and if so, if variable stoma sizes have proportional effects on the water relations of the plant. It would also be of great interest if the described changes in transcriptional activity are the same in different populations of polyploids derived from separate polyploidization events, or if they are stochastic changes. If changes in gene expression levels of specific genes in multiple separately induced polyploids could directly be linked to physiological or morphological changes in polyploids compared with their parents, this would indicate that genes exist, which are directly and consistently regulated by genome doubling. In this context, it is interesting that recent genome wide studies using microarray technology in *Arabidopsis* (Wang *et al.*, 2006; Pignatta *et al.*, 2010; Yu *et al.*, 2010), maize (Riddle *et al.*, 2010), and potato (Stupar *et al.*, 2007) have reported numerous transcriptional changes between ploidy levels, but did not uncover genes that were specifically ploidy-regulated or could have easily explained morphological differences between genotypes of different ploidy levels. Li *et al.* (2012) argue that maybe subtle changes below the detection limit of the array-based experiments, yet discernable by quantitative PCR, could contribute significantly to the observed morphological variation.

In contrast to the studies on autopolyploids cited in the previous section, recent molecular approaches from several labs have shown that allopolyploids display significant variation in gene expression compared with that of their parents (Adams *et al.*, 2003; Wang *et al.*, 2006; Flagel *et al.*, 2008; Chaudhary *et al.*, 2009; Buggs *et al.*, 2010; Chelaifa *et al.*, 2010), but in many cases unequivocal connections between the change in gene expression and an adaptive role in stress responses for the observed transcriptional change have not yet been made. A pair of recent studies in *Coffea* (coffee) addressed the question whether an allopolyploid compared with its parent species had a broader range of transcriptional expression patterns, and thus theoretically increased its chances to be better at responding to changing environmental conditions (Bardil *et al.*, 2011; Combes *et al.*, 2012). In the first of the two studies, the authors used microarray analysis to compare gene expression in *Coffea arabica* and its two progenitor species *C. eugenioides* and *C. canephora*. Their analysis showed that homoeologous gene pairs were differentially regulated in response to temperature in *C. arabica* (Bardil *et al.*, 2011). In a separate study, the authors analyzed the relative contribution of each gene in homoeologous gene pairs to the overall expression of the gene in the allopolyploid and reported significant differences in their relative contribution to the overall transcription of the pair, depending on both tissue type and environmental conditions (Combes *et al.*, 2012). This study suggested that the allopolyploid has a different, but not necessarily broader, range of responses to changing conditions. It would be interesting to see if the difference in expression could lead to a selective advantage, for example, in a comparative study of whole-plant performance of different individuals in different environments.

Homoeologous genes in the allopolyploid *Gossypium hirsutum* showed differential expression levels of each homoeolog when a variety of abiotic stresses, such as heat or cold, were applied (Liu and Adams, 2007; Dong and Adams, 2011). Also, in the allopolyploid *Brassica napus,* stressful environmental conditions led to a change in the identity of splice variants in a significant subset of homoeologous genes when compared to the two parent species *B. olaracea* and *B. rapa* (Zhouet *al.*, 2011). Alternative splicing creates multiple mature mRNAs from a single gene, which can give rise to multiple proteins with different functions. Collectively, these studies provide strong experimental evidence that regulation of gene expression in homoeologs of allopolyploids can be variable in different conditions. Whether such changes in homoeolog regulation are adaptive, again, remains to be shown.

Another example for changed responsiveness to stress of allopolyploids relative to its parents comes from *Arabidopsis*. Here, mRNA stability was compared between allopolyploids and their parent species (Kim and Chen, 2011). Results from this study showed that genes involved in abiotic and biotic stress responses were significantly more likely to also have a differential mRNA decay time when comparing allopolyploids and progenitor species (Wang *et al.*, 2006; Kim and Chen, 2011). These studies suggest that differential post-transcriptional regulation between allopolyploids and their progenitors might modulate the stress response of allopolyploids relative to their parent species and potentially allow survival under altered conditions.

Taken together, modern molecular studies suggest that the duplicated homoeologous genes in allopolyploids might be undergoing subfunctionalization, allowing the allopolyploid to differentially use the homoeologs for different or variable responses to an array of stressful conditions. This in turn would support the idea that allopolyploidization can lead to a functional increase in response options to a wider spectrum of potentially stressful environmental conditions, which could be evolutionarily adaptive. It is important to note, however, that in many cases we still lack direct evidence that polyploidy-induced variation is necessarily linked to the evolutionary potential of populations of polyploids. To gain answers to the question whether or not polyploidy promotes evolution via increased adaptability, more work is needed that integrates molecular and ecological approaches to test the role of specific genes on adaptation under field or controlled conditions.

CONCLUDING REMARKS

Despite the widespread occurrence of polyploidy in nature and the prevalence of its footprint in all angiosperm genomes, the question to what degree polyploidy affects the evolutionary trajectory of a species is still unclear. Old questions about the role of polyploidy in the response to stress or whether genome duplication is beneficial or detrimental to evolutionary success are being revisited using modern genomic tools.

Molecular studies have provided evidence for genomic change on numerous levels of regulation concomitant to polyploidization. However, in many cases we still don't know the effect of polyploidy on fitness under different environmental conditions, and there still is little evidence that observed transcriptional and genomic changes actually lead to faster evolution or greater adaptation in natural populations. Moreover, recent phylogenetic studies perhaps point to the contrary. It is possible that inherent differences in the types of responses to polyploidization are either species-specific or stochastic in individuals. Such differences may be the reason for why some polyploids, such as cotton and soybean, have been so successful and why, as described by Mayrose *et al.*(2011), so many others are not. To answer the question of the role of polyploidy per se on speciation, it would be valuable if future phylogenetic studies could distinguish between autopolyploids and allopolyploids so that the question of genome doubling and genome merger can be disentangled. Integrative approaches that combine molecular and ecophysiological tools that compare responses of polyploids grown in controlled conditions with those grown in natural environments might help in the future to get answers to some of the still unresolved questions, and help us better understand the ecological and evolutionary role that polyploidy has.

ACKNOWLEDGEMENTS

I thank Barbara Mable, Alexa Tullis, Cristy Walcher, members of my lab, and two anonymous reviewers for critical reading of the manuscript and suggestions for improvement. This article is based upon work in my lab and supported by the National Science Foundation under Grant No. 1118539 (to AM).

REFERENCES

1. Adams K, Wendel J (2005). Polyploidy and genome evolution in plants. *Curr Opin Plant Biol* 8: 135–141.
2. Adams KL, Cronn R, Percifield R, Wendel JF (2003). Genes duplicated by polyploidy show unequal contributions to the transcriptome and organ-specific reciprocal silencing. *Proc Natl Acad Sci USA* 100: 4649–4654.
3. Arrigo N, Barker MS (2012). Rarely successful polyploids and their legacy in plant genomes. *Curr Opin Plant Biol* 15: 140–146.
4. Bardil A, de Almeida JD, Combes MC, Lashermes P, Bertrand B (2011). Genomic expression dominance in the natural allopolyploid Coffea arabica is massively affected by growth temperature. *New Phytol* 192: 760–774.
5. Birchler JA, Yao H, Chudalayandi S, Vaiman D, Veitia RA (2010). Heterosis.*The Plant Cell Online* 22: 2105–2112.
6. Blanc G, Wolfe KH (2004). Widespread paleopolyploidy in model plant species inferred from age distributions of duplicate genes. *Plant Cell* 16: 1667–1678.
7. Buggs RJ, Elliott NM, Zhang L, Koh J, Viccini LF, Soltis DE *et al.* (2010). Tissue-specific silencing of homoeologs in natural populations of the recent allopolyploid Tragopogon mirus. *New Phytol* 186: 175–183.
8. Buggs RJA, Zhang L, Miles N, Tate JA, Gao L, Wei W *et al.* (2011). Transcriptomic shock generates evolutionary novelty in a newly formed, natural allopolyploid plant. *Curr Biol* 21: 551–556.
9. Chaudhary B, Flagel L, Stupar RM, Udall JA, Verma N, Springer NM *et al.* (2009). Reciprocal silencing, transcriptional bias

and functional divergence of homeologs in polyploid cotton (gossypium). *Genetics* 182: 503–517.

10. Chelaifa H, Monnier A, Ainouche M (2010). Transcriptomic changes following recent natural hybridization and allopolyploidy in the salt marsh species Spartina × townsendii and Spartina anglica (Poaceae). *New Phytol* 186: 161–174.
11. Chen Z (2007). Genetic and epigenetic mechanisms for gene expression and phenotypic variation in plant polyploids. *Annu Rev Plant Biol* 58: 377–406.
12. Chen ZJ (2010). Molecular mechanisms of polyploidy and hybrid vigor.*Trends Plant Sci* 15: 57–71.
13. Comai L (2005). The advantages and disadvantages of being polyploid. *Nat Rev Genet* 6: 836–846.
14. Comai L, Tyagi AP, Winter K, Holmes-Davis R, Reynolds SH, Stevens Y *et al.* (2000). Phenotypic instability and rapid gene silencing in newly formed*Arabidopsis* allotetraploids. *Plant Cell* 12: 1551–1568.
15. Combes M-C, Cenci A, Hln Baraille, Bertrand BÆ, Lashermes P (2012). Homeologous gene expression in response to growing temperature in a recent allopolyploid (Coffea arabica L.). *J Hered* 103: 36–46.
16. Coyne J, Orr H (2004) *Speciation*. Sinauer Associates, Inc: Sunderland, MA.
17. Dong S, Adams KL (2011). Differential contributions to the transcriptome of duplicated genes in response to abiotic stresses in natural and synthetic polyploids. *New Phytol* 190: 1045–1057.
18. Doyle JJ, Doyle JL, Brown AHD (1999). Origins, colonization, and lineage recombination in a widespread perennial soybean polyploid complex. *Proc Natl Acad Sci* 96: 10741–10745.
19. Doyle JJ, Doyle JL, Rauscher JT, Brown AHD (2004). Evolution of the perennial soybean polyploid complex (Glycine subgenus Glycine): a study of contrasts. *Biol J Linnean Soc* 82: 583–597.
20. Flagel L, Udall J, Nettleton D, Wendel J (2008). Duplicate gene expression in allopolyploid Gossypium reveals two temporally distinct modes of expression evolution. *BMC Biol* 6: 16.
21. Fryxell PA (1979) *The Natural History of the Cotton Tribe*. Texas A&M University Press: College Station, TX, USA.

22. Gaeta RT, Pires JC, Iniguez-Luy F, Leon E, Osborn TC (2007). Genomic changes in resynthesized *Brassica napus* and their effect on gene expression and phenotype. *Plant Cell* 19: 3403–3417.
23. Gaeta RT, Yoo S-Y, Pires JC, Doerge RW, Chen ZJ, Osborn TC (2009). Analysis of gene expression in resynthesized *Brassica napus* allopolyploids using arabidopsis 70mer oligo microarrays. *PLoS ONE* 4: e4760.
24. Grant V (1971) *Plant Speciation,* 1st edn. Columbia University Press: New York.
25. Grant V (1975) *Genetics of flowering plants*. Columbia University Press: New York.
26. Gu Z, Steinmetz LM, Gu X, Scharfe C, Davis RW, Li W-H (2003). Role of duplicate genes in genetic robustness against null mutations. *Nature* 421: 63–66.
27. Gustavsson A (1946). Apomixis in higher plants. *Lunds Universitets Arsskrift*42–43: 1–370.
28. Hagerup O (1932). Über polyploidie in beziehung zu klima, ökologie und phylogenie. *Hereditas* 16: 19–40.
29. Hanelt P (1966). Polyploidie-frequenz und geographische verbreitung bei höheren pflanzen. *Biologische Rundschau* 4: 183–196.
30. Hegarty M, Barker G, Wilson I, Abbott R, Edwards K, Hiscock S (2006). Transcriptome shock after interspecific hybridization in Senecio is ameliorated by genome duplication. *Curr Biol* 16: 1652–1659.
31. Hegarty MJ, Batstone TOM, Barker GL, Edwards KJ, Abbott RJ, Hiscock SJ (2011). Nonadditive changes to cytosine methylation as a consequence of hybridization and genome duplication in Senecio (Asteraceae). *Mol Ecol* 20: 105–113.
32. Hegarty MJ, Jones JM, Wilson ID, Barker GL, Coghill JA, Sanchez-Baracaldo P*et al.* (2005). Development of anonymous cDNA microarrays to study changes to the Senecio floral transcriptome during hybrid speciation. *Mol Ecol* 14: 2493–2510.
33. Hodgson JG, Sharafi M, Jalili A, Diaz S, Montserrat-Marti G, Palmer C *et al.* (2010). Stomatal vs. genome size in angiosperms: the somatic tail wagging the genomic dog? *Ann Bot (Lond)* 105: 573–584.

34. Jiao Y, Wickett NJ, Ayyampalayam S, Chanderbali AS, Landherr L, Ralph PE*et al*. (2011). Ancestral polyploidy in seed plants and angiosperms. *Nature*473: 97–100.
35. Johnson HB (1975). Plant pubescence: an ecological perspective. *Bot Rev*41: 233–258. |
36. Kim E-D, Chen ZJ (2011). Unstable transcripts in arabidopsis allotetraploids are associated with nonadditive gene expression in response to abiotic and biotic stresses. *PLoS ONE* 6: e24251.
37. Lee HS, Chen ZJ (2001). Protein-coding genes are epigenetically regulated in Arabidopsis polyploids. *Proc Natl Acad Sci USA* 98: 6753–6758.
38. Levin D (1975). Minority cytotype exclusion in local plant populations. *Taxon*24: 35–43.
39. Levin D (2002) *The Role of Chromosomal Change in Plant Evolution*. Oxford University Press: New York.
40. Lewis W (1980). Polyploidy in species populations. In: Lewis W (ed.).*Polyploidy: Biological Relevance*. Plenum: New York. pp 103–144.
41. Li X, Yu E, Fan C, Zhang C, Fu T, Zhou Y (2012). Developmental, cytological and transcriptional analysis of autotetraploid *Arabidopsis*. *Planta* 236: 579–596.
42. Li W-L, Berlyn GP, Ashton PMS (1996). Polyploids and their structural and physiological characteristics relative to water deficit in Betula papyrifera (Betulaceae). *Am J Bot* 83: 15–20.
43. Liu Z, Adams KL (2007). Expression partitioning between genes duplicated by polyploidy under abiotic stress and during organ development. *Curr Biol*17: 1669–1674.
44. Lynch M (2007) *The Origins of Genome Architecture*. Sinauer: Sunderland, MA.
45. Löve A (1953). Subarctic polyploidy. *Hereditas* 39: 113–124.
46. Löve A, Löve D (1949). The geobotanical significance of polyploidy.*Portugaliae Acta* (Suppl) 273–352.
47. Madlung A, Henkhaus N, Jurevic L, Kahsai EA, Bernhard J (2012). Natural variation and persistent developmental instabilities in geographically diverse accessions of the allopolyploid Arabidopsis suecica. *Physiol Plant* 144: 123–133.

48. Madlung A, Tyagi AP, Watson B, Jiang H, Kagochi T, Doerge RW *et al.* (2005). Genomic changes in synthetic *Arabidopsis* polyploids. *Plant J* 41: 221–230.
49. Maherali H, Walden AE, Husband BC (2009). Genome duplication and the evolution of physiological responses to water stress. *New Phytol* 184: 721–731.
50. Masterson J (1994). Stomatal size in fossil plants: evidence for polyploidy in majority of angiosperms. *Science* 264: 421–424.
51. Matsushita SC, Tyagi AP, Thornton GM, Pires JC, Madlung A (2012). Allopolyploidization lays the foundation for evolution of distinct populations: evidence from analysis of synthetic arabidopsis allohexaploids. *Genetics*191: 535–547.
52. Mayfield D, Chen ZJ, Pires JC (2011). Epigenetic regulation of flowering time in polyploids. *Curr Opin Plant Biol* 14: 174–178.
53. Mayr E (1963) *Animal Species and Evolution*. Harvard University Press: Cambridge.
54. Mayrose I, Zhan SH, Rothfels CJ, Magnuson-Ford K, Barker MS, Rieseberg LH*et al.* (2011). Recently formed polyploid plants diversify at lower rates.*Science* 333: 1257.
55. Melaragno JE, Mehrotra B, Coleman AW (1993). Relationship between endopolyploidy and cell size in epidermal tissue of arabidopsis. *The Plant Cell Online* 5: 1661–1668.
56. Moore RC, Purugganan MD (2005). The evolutionary dynamics of plant duplicate genes. *Curr Opin Plant Biol* 8: 122–128.
57. Müntzing A (1936). The evolutionary significance of autopolyploidy.*Hereditas* 21: 363–378.
58. Ohno S (1970) *Evolution by Gene Duplication*. Springer Verlag: New York.
59. Otto SP (2007). The evolutionary consequences of polyploidy. *Cell* 131: 452–462. |
60. Pignatta D, Dilkes BP, Yoo SY, Henry IM, Madlung A, Doerge RW *et al.* (2010). Differential sensitivity of the Arabidopsis thaliana transcriptome and enhancers to the effects of genome doubling. *New Phytol* 186: 194–206.
61. Pumphrey M, Bai J, Laudencia-Chingcuanco D, Anderson O, Gill BS (2009). Nonadditive expression of homoeologous genes is established upon polyploidization in hexaploid wheat. *Genetics* 181: 1147–1157.

62. Ramsey J (2011). Polyploidy and ecological adaptation in wild yarrow. *Proc Natl Acad Sci* 108: 7096–7101.
63. Ramsey J, Schemske DW (1998). Pathways, mechanisms, and rates of polyploid formation in flowering plants. *Annu Rev Ecol Syst* 29: 467–501.
64. Rapp R, Udall J, Wendel J (2009). Genomic expression dominance in allopolyploids. *BMC Biol* 7: 18.
65. Rhoades MM, Dempsey E (1966). Induction of chromosome doubling at meiosis by the elongate gene in maize. *Genetics* 54: 505–522.
66. Riddle N, Jiang H, An L, Doerge R, Birchler J (2010). Gene expression analysis at the intersection of ploidy and hybridity in maize. *TAG Theor Appl Genet*120: 341–353.
67. Riddle NC, Kato A, Birchler J (2006). Genetic variation for the response to ploidy change in Zea mays L. *Theor Appl Genet* 114: 101–111.
68. Schlueter JA, Dixon P, Granger C, Grant D, Clark L, Doyle JJ *et al.* (2004). Mining EST databases to resolve evolutionary events in major crop species.*Genome* 47: 868–876.
69. Soltis DE, Buggs RJA, Doyle JJ, Soltis PS (2010). What we still don't know about polyploidy. *Taxon* 59: 1387–1403.
70. Soltis P, Soltis D (2009). The role of hybridization in plant speciation. *Ann Rev Plant Biol* 60: 561–588.
71. Speckman G, Post J, Dijkstra H (1965). Length of stomata as an indicator for polyploidy in rye-grasses. *Euphytica* 14: 225–228.
72. Stebbins GJ (1950) *Variation and Evolution in Plants*. Columbia University Press: New York.
73. Stebbins GL (1947). Types of polyploids: their classification and significance. *Adv Genet* 1: 403–429.
74. Stupar RM, Bhaskar PB, Yandell BS, Rensink WA, Hart AL, Ouyang S *et al.* (2007). Phenotypic and transcriptomic changes associated with potato autopolyploidization. *Genetics* 176: 2055–2067.
75. Wang J, Tian L, Lee HS, Wei NE, Jiang H, Watson B *et al.* (2006). Genomewide nonadditive gene regulation in Arabidopsis allotetraploids.*Genetics* 172: 507–517.

76. Wang J, Tian L, Madlung A, Lee HS, Chen M, Lee JJ *et al.* (2004). Stochastic and epigenetic changes of gene expression in Arabidopsis polyploids. *Genetics* 167: 1961–1973.

77. Wendel JF, Cronn RC (2003). Polyploidy and the evolutionary history of cotton. *Adv Agron* 78: 139–186.

78. Winge O (1917). The chromosomes, their number and general importance.*Compt Rend Trav Lab Carlsberg* 13: 131–275.

79. Wood T, Takebayashi N, Barker M, Mayrose I, Greenspoon P, Rieseberg L (2009). The frequency of polyploid speciation in vascular plants. *Proc Natl Acad Sci USA* 106: 13875–13879.

80. Wright KM, Pires JC, Madlung A (2009). Mitotic instability in resynthesized and natural polyploids of the genus *Arabidopsis* (Brassicaceae). *Am J Bot*96: 1656–1664.

81. Xiong Z, Gaeta RT, Pires JC (2011). Homoeologous shuffling and chromosome compensation maintain genome balance in resynthesized allopolyploid *Brassica napus*. *Proc Natl Acad Sci USA* 108: 7908–7913.

82. Yu Z, Haberer G, Matthes M, Rattei T, Mayer KFX, Gierl A *et al.* (2010). Impact of natural genetic variation on the transcriptome of autotetraploid*Arabidopsis thaliana*. *Proc Natl Acad Sci* 107: 17809–17814.

Citations

CHAPTER 1

Jared Diamond, Evolution, consequences and future of plant and animal domestication, doi: 10.1038/nature01019.

CHAPTER 2

Jacob A. Tennessen, Rajanikanth Govindarajulu, Tia-Lynn Ashman, and Aaron Liston, Evolutionary Origins and Dynamics of Octoploid Strawberry Subgenomes Revealed by Dense Targeted Capture Linkage Maps, doi: 10.1093/gbe/evu261.

CHAPTER 3

Bhupendra Chaudhary, "Plant Domestication and Resistance to Herbivory," International Journal of Plant Genomics, vol. 2013, Article ID 572784, 14 pages, 2013. doi:10.1155/2013/572784.

CHAPTER 4

N E G Forsberg, J Russell, M Macaulay, M W Leino, and J Hagenblad, Farmers Without Borders-Genetic Structuring in Century Old Barley (Hordeum Vulgare), doi:10.1038/hdy.2014.83.

CHAPTER 5

Loretta Gratani, "Plant Phenotypic Plasticity in Response to Environmental Factors," Advances in Botany, vol. 2014, Article ID 208747, 17 pages, 2014. doi:10.1155/2014/208747.

CHAPTER 6

John Gottula, Ramsey Lewis, Seiya Saito, and Marc Fuchs, Allopolyploidy and the Evolution of Plant Virus Resistance, doi: 10.1186/1471-2148-14-149.

CHAPTER 7

C C Vigueira, K M Olsen, and A L Caicedo, The Red Queen in the Corn: Agricultural Weeds As Models of Rapid Adaptive Evolution, doi:10.1038/hdy.2012.104.

CHAPTER 8

A Madlung, Polyploidy and its Effect on Evolutionary Success: Old Questions Revisited with New Tools, doi:10.1038/hdy.2012.79.

Index